BEYOND T

John Boslough was educated at Princeton University, where he received honours in the history of science. An award-winning journalist, his articles have appeared in *Smithsonian*, *Science 81*, *Reader's Digest*, *Psychology Today*, the *Washington Post* and elsewhere. He also has written for Associated Press and *National Geographic*, and was Science Editor of *U.S. News & World Report*. He divides his time between Colorado and Virginia.

BEYOND
THE BLACK HOLE

Stephen Hawking's Universe

JOHN BOSLOUGH

Fontana/Collins

First published in Great Britain by William Collins 1985
First issued in Fontana Paperbacks 1986

Copyright © John Boslough 1984

Made and printed in Great Britain by
William Collins Sons & Co. Ltd, Glasgow

CONTENTS

ILLUSTRATIONS

———◆———

We dance round in a ring and suppose,
But the Secret sits in the middle and knows.

ROBERT FROST

ACKNOWLEDGMENTS

———◆———

There are several people who were instrumental in the preparation of this book. I would especially like to thank Stephen Hawking for the many hours he spent with me in Cambridge, his illumination of the manuscript. Anne K. DuVivier's enthusiasm and meticulous gathering of photographs were also essential to the project.

In addition I would like to thank Murray Gell-Mann for taking time off from mushroom-hunting in Aspen, Colorado, to speak with me about the relationship between quantum mechanics and relativity physics. I appreciate, too, the insights offered by Stephen Hawking's colleagues and former colleagues, Ian Moss, Malcolm Perry and Roger Penrose.

Katherine Boslough and James Boslough deserve special thanks for their support and understanding as does Jodi Cobb for her encouragement. Thanks also to Dan O'Keefe, John Brockman, Katinka Matson, Deborah Elise White and Douglas Stumpf as well as Linda Creighton and Stewart Powell for putting up with me on several trips to England.

PROLOGUE

On a spring morning in 1974, a young man dressed in a suit was carried up the steps of a white-colonnaded mansion overlooking London's St James's Park. Placed in a wheelchair inside the building, No. 6 Carlton House Terrace, he was wheeled into a large meeting room to receive one of the highest honours in Great Britain: induction into the Royal Society, among the world's most eminent scientific bodies.

At thirty-two, Stephen William Hawking was one of the youngest inductees in Society history, an honour bestowed for his work in theoretical physics. Tradition dating from the seventeenth century called for newly elected fellows to walk to the podium to shake the president's hand and sign the roll of honour. But at this investiture Sir Alan Hodgkin, Nobel Prize-winning biologist and president of the Society, brought the roll book down from the stage to Hawking's wheelchair at the front of the room. As the new member laboured over his signature, there was protracted silence. When he finished with a broad smile, thunderous applause broke out.

I met Hawking for the first time seven years later in a corridor outside the meeting hall where the investiture had taken place. We were introduced by Roger Penrose, a mathematician and theoretical physicist at Oxford University. Penrose, an old friend and collaborator of Hawking, had himself been inducted into the Society just two years before, partly for work the two men had done together.

Since 1962 Hawking has suffered from a wasting illness called motor neurone disease. It has slowly taken most of

9

his nervous and muscular functions. He cannot walk and can barely talk. I had been warned by Penrose and others that I would find his condition worse than I expected.

In fact, I was stunned. Before me, slumped in his wheelchair, was one of the world's foremost scientists, a man not much older than myself. I estimated his weight at no more than 110 pounds. Because he was so thin, it was impossible to judge his height, although he appeared about average – perhaps five feet nine. His face was youthful, but his body had the frailty and muscle structure of a bedridden old man.

When Penrose had concluded the introductions, Hawking began speaking in such a low voice that I had to bend down to hear. He seemed to be struggling to speak, his voice a laboured moan punctuated with gasps. I looked to Penrose for guidance. He quickly translated Hawking's remark: 'I'll see you in my office at eleven o'clock next Tuesday morning.'

Afterwards I asked Penrose if Hawking had been having a particularly bad day. On the contrary, Penrose said. He thought Hawking looked particularly good.

Every time I saw him I wondered how he managed. He has not walked for over twelve years, and his voice is so feeble that only a few intimates can understand him, yet he has made some of the most significant strides in theoretical physics in his generation, changing the way we look at the universe.

As I came to know Hawking, the truth became apparent. His accomplishment is not due simply to his will to live or to the fact that he is a survivor, though he is certainly a tough and stubborn man. He succeeds because of his intellect, and as the ravages of his disease have, over two decades, taken his physical powers from him, he has come to live a life of the mind.

Hawking's mind is his most powerful tool. It is also his work, his plaything, his recreation, his joy – his life. His wheelchair gives him a special vantage point for the major preoccupation of that mind: the universe we inhabit, how it came into being, how it operates, and how it will end. A totally cerebral man, he demonstrates the power of the human intellect to fathom the universe when a restless mind is set free.

CHAPTER ONE

❖

Quarks and Quasars

'It is the most persistent and greatest adventure in human history, this search to understand the universe, how it works and where it came from. It is difficult to imagine that a handful of residents of a small planet circling an insignificant star in a small galaxy have as their aim a complete understanding of the entire universe, a small speck of creation truly believing it is capable of comprehending the whole.'

Murray Gell-Mann, who made this statement, is one of a group of theoretical physicists engaged on this adventure. They are searching for a single interaction at the heart of the universe, one that will explain all of the phenomena that surround us.

The task of finding this single interaction is so monumental that it eluded even Einstein, who spent the last thirty years of his life in an unsuccessful search for unity. We are a little closer today, nearly thirty years after Einstein's death, but still the universe seems to operate by several sets of rules that act in layers, independent of each other.

The most apparent of these basic rules of nature, gravity, controls the biggest objects in the universe – the stars, the planets, you and me. The other three rules that scientists have uncovered operate at the subatomic level: the strong nuclear force, trillions of times more powerful than gravitation, holds the nucleus of an atom together; electromagnetism keeps electrons in place around the nucleus, making ordinary matter seem solid; the weak nuclear force causes radioactive decay in certain atoms like uranium.

Groping in a morass of mathematics, Einstein was unable to reconcile these different sets of natural laws. He believed in his heart that beyond them lay a final simplicity in which they could all be explained as a single law. This belief was based purely on an aesthetic appeal, a notion of an irreducible set of equations that could explain everything.

Not all physicists believe that such unification is possible. Wolfgang Pauli, an Austrian theorist, once joked, 'What God hath put asunder, no man shall ever join.' But a unified theory is not something science actually requires for its continued progress. Physicists need a unified theory only in the sense that Sir Edmund Hillary needed to climb Mount Everest.

If such a law is found, it could prove almost meaningless, or it could lead to a new golden age of science. Scientists don't know, just as they had no idea that Einstein's unification of mass and energy would lead to the age of the atom. Or that quantum mechanics, the mathematical system used by physicists to explain the movement of subatomic particles, would be used to make the first laser. Yet a unification theory remains a religious vision to some scientists, a Zen-like view of reality in which all the forces and all the matter in nature come from a single source.

If you look at the world around you today, a reconciliation of such diverse forces appears far from possible. The reason is that we live in a low-energy, cold universe, one in which forces and matter seem stable and disconnected. But the universe was not always as we observe it today. The cosmos has cooled down dramatically since the moment of its origin. As it cooled, the infant universe left a trail of clues that physicists have followed back to the beginning. Most physicists believe the key to the universe lies there, at the instant of the Big Bang or shortly thereafter. In that instant, the four forces may have existed in the intense energy of the primordial cataclysm for a fraction of a second as a single interaction. This interaction is thought to be so basic that all subsequent forces have descended from it.

Theoretical physicists using the latest mathematical reconstructions have developed a good idea of what happened within less than a billion trillionth of a second after the Big Bang. A remarkable achievement, but still it doesn't go far enough back in time for them to see, in their equations, the moment when all the forces and laws of nature were unified.

In later stages each of the four forces has had a time of dominance in the history of the universe, like periods of ascendancy of political parties in democratic governments. In the universe we inhabit, gravitation, the weakest but most pervasive, is the major force. Its pull acts over vast distances – on galaxies, stars and quasars, the most distant and least understood objects in the universe. Gravitation has been the major factor during almost all of the universe's fourteen or fifteen-billion-year lifetime. Before that, in the first few seconds after the Big Bang, the weak nuclear force prevailed, and before that electromagnetism.

It is likely that the strong nuclear force was almost completely dominant during the first few billionths of a second after the Big Bang, an instant when matter and energy were one, and stars and galaxies had not yet developed. In the billionths of a second earlier in the history of the cosmos, energy was so intense that none of the four forces could be distinguished from the others. At least, most theorists have convinced themselves of that scenario.

'It is the job of the theoretical physicist, using all the mathematical tools at his disposal, to find out what happened before things cooled down enough for the four forces to divide and obscure the underlying interaction,' Sheldon Glashow, a Harvard theorist, told me on a rainy day in August 1982 at the Aspen Physics Centre in

Colorado. 'A lot of people, including myself, are working on that very problem. But nobody has yet shown that all the interactions were in fact one and the same in the very early universe.'

Glashow has led the way in the search for the underlying interaction. During the 1960s he tried – without success – to group certain short-lived subatomic particles in ways that would lead to this unifying force. His approach kept producing unexplained and unworkable mathematical infinities.

Steven Weinberg, then at the Massachusetts Institute of Technology, and Abdus Salam, at Imperial College in London, were more successful. Working independently, they produced in 1967 a set of equations that seemed to prove that the weak nuclear force and electromagnetism were, if certain obscuring factors were ignored, one and the same.

The beauty of the Weinberg–Salam model was that it predicted that certain events would occur under special conditions in particle accelerators, the atom smashers that physicists use to strip away the many layers of atoms. Weinberg, Salam and Glashow shared the Nobel Prize in 1979 for this work.

During the 1970s other physicists developed different groups of calculations that purported to show that not only were the weak force and electromagnetism the same, but also that the strong force holding the nuclei of atoms together was a member of the same family. These types of calculation are called grand unified theories, or GUTs.

Some scientists are not so sure the GUTs approach is exactly on target. Murray Gell-Mann says, 'They are neither grand, nor unified. It might even be said that they are not even theories – just glorified models.' Still, he admits that the approach may be one of the most promising in the pursuit of the underlying interaction.

Gell-Mann himself originated the concept of quarks, the sub-subatomic particles that most theorists believe are the fundamental constituents of the protons and neutrons that make up the nuclei of every atom in the universe. Before Gell-Mann conceived and named them (indirectly from a line in James Joyce's *Finnegans Wake,* 'Three quarks for Muster Mark'), particle physics was in a state of disarray, having failed miserably to cope with the dozens of new particles found in accelerators in the 1950s and early 1960s. As a result of Gell-Mann's quark synthesis, particle physicists once again viewed the atom's core as a more or less orderly little universe of its own.

Gell-Mann admits he would like to see a unification of the four forces, but he's not sure it will happen in his lifetime. 'Nobody's even yet shown that the three forces at work inside the atom have the same root. Some people may be close. I don't know. But it hasn't been shown to me yet.'

And what about gravity, the force we are all most familiar with? Where does it fit into grand unification? Although particle physicists may be coming close to a unified theory of the universe with the three forces that push and pull within the atom, gravitation is still the odd force out. And this despite the fact that the vast world of cosmology and the tiny universe within the atom are finally converging as particle physicists looking inward with their giant accelerators and cosmologists looking outward with their telescopes begin to realize they are looking at the same thing.

There are several groups of scientists working on unifying all four forces, trying to add gravity to the other three. Gell-Mann told me, 'Most of them don't know what they're doing. They're just using various mathematical tricks.' He conceded somewhat cautiously that there was one group of theorists who had a chance of making some progress towards finding this great secret of the universe.

The group is headed by Stephen Hawking at Cambridge University. 'Hawking is the only one on the relativity side who understands particle physics,' said Gell-Mann. 'He's a remarkable man, an absolutely astonishing fellow.'

CHAPTER TWO

Against the Odds

Stephen Hawking was the eldest of four children of a bookish, tightly knit family. His father was a research biologist in tropical diseases for the National Institute of Medical Research. Born on 8 January 1942 in Oxford, he grew up in London and in the city of St Albans, about twenty miles to the north. From the age of eleven he attended St Albans Public School, a school his parents hoped would prepare him for entrance into Oxford University.

By the time he was eight or nine, he knew that he wanted to be a scientist. He had already shown a knack for taking clocks and radios apart to find out how they worked, and science seemed to him to be where the truth about the things surrounding him would be found. As a teenager, though, he found much of science too imprecise: 'The biological sciences were too descriptive, too hazy for me,' Hawking recalls. 'Of course, they have become more exact today because of molecular biology.' By the time he was fourteen, he had made up his mind to become a mathematician or physicist. Fearing his son would never find work, Hawking's father tried unsuccessfully to talk him out of it.

At about the same time, Hawking's mind took a sceptical turn. When he was fifteen, he tried the same dice-throwing experiments being conducted in the extrasensory programme at Duke University in the 1950s. After following the Duke experiments closely for a while, he became convinced that ESP was a fraud. 'Whenever the experiments got results, the experimental techniques were faulty,' he says now. 'Whenever the experimental techniques were sound, the results were no good.'

To this day he thinks parapsychology is a waste of time. 'People taking it seriously are at the stage where I was when I was a teenager,' he says, laughing.

Despite these occasional bouts of precocity, he was not outstanding in secondary school. His parents worried he might fail his entrance exam for Oxford, and his father, an alumnus of University College, tried to pull strings to ensure his acceptance. But the father had underestimated the son. Stephen received nearly perfect marks on the physics section of the entrance exams, and performed so well during the interview that there was no question about admission; he entered Oxford in 1959.

At Oxford, Hawking was a popular student, known for his wit, and at one time the coxswain for one of the college's eight-man rowing shells. Most of the people who remember him from those days recall a spirited undergraduate with long hair and an interest in classical music and science fiction. He took an independent and freewheeling approach to studies, although his tutor, Dr Robert Berman, recalls that he and other dons were aware that Hawking had a first-rate mind, 'completely different from his contemporaries'.

He was so good at physics that he had to put little work into it. 'Undergraduate physics was simply not a challenge for him,' says Berman. 'He could do any problem put before him without even trying.' One day in class, after reading a solution he had worked out, he screwed the paper up and disdainfully threw it across the room into a wastepaper basket.

Hawking could prove forgetful at crucial – or convenient – moments. During his last year at Oxford, he applied for a job with the Ministry of Works. He then forgot to appear for the examination. Had he passed, he might well have ended up taking care of monuments.

When the time came to graduate, Hawking needed first-class honours to receive a scholarship for a graduate physics course at Cambridge. At a crucial oral exam he responded thus to an examiner about his plans: 'If I get a first, I shall go to Cambridge. If I receive a second, I will remain at Oxford. So I expect that you will give me a first.' Those who know him agree it's pure Hawking.

Dr Berman reported later of Hawking's meeting with the examiners: 'At least they were intelligent enough to realize they were talking to someone more intelligent than they.' Hawking received his first and started the graduate course at Cambridge the following year.

By then Hawking had settled on a career in theoretical physics specializing in cosmology. He had considered other areas in physics, but only briefly. Once, while taking a special summer course at the Royal Greenwich Observatory, he helped Sir Richard Woolley, then Astronomer Royal, to measure the constituents of a double star. When he looked through the observatory's telescope, he was profoundly disappointed to see just a pair of fuzzy spots of light going in and out of focus.

Since then he has only looked through a telescope once or twice and has remained unimpressed with observational astronomy. Theory was always more exciting for Hawking, and cosmology the most exciting of all, since it involves the question, Where does the universe come from?

By the time he was a graduate he had begun showing signs of becoming a top theoretical physicist. Roger Penrose, then a research associate at King's College, London, recalls first coming across Hawking in those days. 'He used to ask the most awkward questions, questions that were very difficult to answer,' Penrose remembers. 'He would always aim right at the weakest part of your argument. But it was not easy then to tell how original he was going to become.'

But signs of serious illness appeared at the beginning of

his first year as a graduate – a lack of dexterity and slight paralysis that made it difficult for Hawking to tie his shoes and, occasionally, to talk. Doctors soon diagnosed his illness as amyotrophic lateral sclerosis, or motor neurone disease, a rare and potentially crippling syndrome. It is sometimes called Lou Gehrig's disease, after the Yankee first baseman who died from it. It also claimed the life of David Niven in 1983.

Motor neurone disease is marked by the gradual disintegration of nerve cells in the spinal cord and brain that regulate voluntary muscular activity. The first symptoms are weakness and twitching of the hands along with, perhaps, slurred speech or difficulty in swallowing. As the neurons stop functioning, the muscles under their control atrophy; a victim becomes increasingly disabled although the mind remains lucid. Death usually occurs either from pneumonia or suffocation, when the respiratory muscles finally fail.

Doctors hoped that Hawking's illness would stabilize, but his condition continued to deteriorate. He was given just a couple of years to live. 'I was understandably quite depressed at the prognosis,' Hawking recalls. The prospect of an early death propelled him into a stupefying depression for two years, a period in which he spent little time on his research and a great deal of time in his room listening to classical music – mostly Wagner – and reading science fiction. He also began 'drinking a fair amount'.

His tutor, a theorist named Dennis Sciama who headed the general relativity group at Cambridge, was both aware of his student's potential and concerned about his illness. 'He always had a feeling for what we were discussing. With other bright students, it might take a couple of years. With Stephen it was just a month. He was always saying, "But . . ." to almost any statement you would make.' Sciama allowed Hawking to indulge his depression. If he wanted to drink himself into a stupor to forget his troubles, fine; if

21

he didn't want to work on his thesis, too bad. But Sciama turned down an appeal from Hawking's father to help his son finish his dissertation early.

As the months passed, Hawking's condition began to stabilize. He realized that death was not imminent. His spirits lifted, and with the encouragement of friends, family and tutor his natural buoyancy reappeared. He also began to realize that he was working in a purely cerebral area — one with virtually no emphasis on human physical prowess. The disease had not affected his mind; it would not affect his work. The depression vanished, Sciama urged him on, and he started to work on his dissertation again.

At about this time, one of the most important events in Hawking's life occurred: he attended a party and met Jane Wilde, a student of languages in London. In 1965, after a two-year courtship carried on between London and Cambridge, they were married. 'He already had the beginnings of the condition when I first knew him, so I've never known a fit, able-bodied Stephen,' she says. 'I simply decided what I was going to do, and I did it.'

Hawking's marriage was the turning point. 'It made me determined to live, to go on. Jane really gave me the will to live.'

Everybody who knows her describes Jane Hawking as a remarkable woman. During the first year of their marriage she commuted between London and Cambridge so that she could finish her own graduate studies, and in the meantime typed her husband's dissertation. For nearly two decades she has taken care of Hawking's physical needs; she also has made certain that the Hawking family leads a relatively normal life, in spite of both Hawking's disability and the fame that has recently come his way. Their first child, Robert, was born in 1967. A daughter, Lucy, arrived three years later, and Timothy was born in 1979.

Although Jane and other people around Stephen are protective to a degree, they all tend to ignore his condition.

'Stephen doesn't make any concessions to his illness, and I don't make any concessions to him,' Jane once said. The major problem in her life is not her husband's physical condition; it is that she cannot follow all the details of his work in theoretical physics.

During the three years after he received his doctorate, Hawking worked as a research associate at Cambridge and began collaborating with Penrose on what was to be his first major piece of research, the mathematical proof of the beginning of time. His physical condition was deteriorating again, and by the early 1970s Hawking was permanently confined to a wheelchair. But by then his mind was soaring. His induction into the Royal Society in 1974 was a stunning triumph for a man who, a decade earlier, thought he would not live to his twenty-fifth birthday.

Those were happy years, both professionally and personally, for Jane and Stephen Hawking. Since then his condition has more or less stabilized, although some of his colleagues believe that it has become more difficult to understand him in the past year or two. And some of his friends, particularly those who do not see him regularly, fear that his overall condition has worsened again over the past few years.

The dirty brick structure housing the Department of Applied Mathematics and Theoretical Physics, where Hawking works, looks like an abandoned nineteenth-century factory lost among the Gothic façades and spires of Cambridge. Its main door faces onto an alleyway off Silver Street. Towards the rear of the building, in another alley, is a twenty-five-foot ramp Hawking uses to enter the building through a swing door. He commutes each day by motorized wheelchair from his home on the ground floor of a Victorian house in West Road about half a mile away.

The office facing into a grey and uninviting lounge is scientific Gothic. It contains racks of physics texts, a computer terminal, pictures of three handsome children, and a special page-turner Hawking fought the bureaucracy to obtain. There is also a specially fitted telephone that now sits idle. Lists of scientific papers are suspended by transparent tape on the walls so he can view them easily.

It is almost impossible to understand Hawking upon first meeting him. After a few hours of listening closely to his thin monotone – translated by Judy Fella, the young woman who was then his secretary – I found I could understand about half of what he was saying. Some words were incomprehensible even to Fella, who had worked with him for years, and Hawking was forced to spell them out. I told him, to his amusement, that part of the problem for an American in understanding him was his British accent.

As he works, his body occasionally droops down into his wheelchair, his head collapsing onto his chest. He has almost no head or facial control, and a smile sometimes turns into a grimace. Nonetheless, when I first appeared in his office, Hawking greeted me with an impish grin, his blue eyes twinkling behind heavy glasses.

His brown hair, flecked with grey, is early Beatles, and he normally dresses in standard scientist garb: baggy trousers, garish tie often mismatched with a broad-striped shirt, plaid or tweed jacket, academic soft soles or boots with bottoms noticeably unused.

Hawking thinks things through carefully before speaking so that he will not have to repeat himself. He does not waste words. Sometimes after he stops working for a few minutes – for a bit of secretarial business or tea – he resumes talking in mid-sentence exactly where he left off. He so completely ignores his physical limitations that, after a while, I found myself doing the same.

One day, as I spoke with him, I had become so oblivious to his condition that I began talking about a problem I was having with my elbow as the result of a squash match in

London the day before. Hawking made no comment. He simply steered his wheelchair out of the room and waited in the hall for me to return to the subject at hand, theoretical physics.

Most days at work Hawking just thinks. He spends much of his time developing new approaches to problems in theoretical physics. One of his colleagues, Ian Moss, told me one morning, 'Stephen comes up with all the ideas. The rest of us only test them out to see if they work.'

Hawking is blessed with a prodigious memory. He is able to work out and retain page after page of complex equations, weaving the mathematical hieroglyphs together as an ordinary person might arrange the words in a sentence. The University of Alberta's Werner Israel, a theoretical physicist and Hawking's co-author on the book *General Relativity*, has said that his feats of memory are akin to Mozart's composing an entire symphony in his head.

His colleagues are often stunned by what Hawking has remembered. A secretary who worked for him while he was visiting the California Institute of Technology said he once recalled twenty-four hours later a tiny mistake he had made while dictating – from memory – forty pages of equations.

One of Hawking's students told me that, while driving him to London for a physics conference once, Hawking remembered the page number of a minute error he had read in a book years before. Other physicists have said that the complex equations that pour forth from his mind are both elegant and inspired – the ultimate accolades for a theoretical physicist.

Hawking's work has drawn an outstanding group of theoretical physicists to Cambridge from both sides of the Atlantic. Most days at lunch and again at teatime they assemble to share their wit and wisdom with Hawking, and the discussion jumps from red shifts and quantum effects to black holes and singularities at the beginning of time, light-years beyond the surroundings.

The talk is fast, punctuated with put-downs and one-liners. 'Hey, Stephen's showing his age,' one graduate student says when Hawking makes a minor mathematical error. Hawking lights up at such lines, and these sessions can be the high point of the day. One of his students once told me that teatime with Stephen can be more enlightening than a term with somebody else.

It is remarkable that Hawking has been able to achieve what he has. Doctors, in fact, believe it is a miracle he is alive. An American doctor familiar with Hawking's disease told me that each day he lives he sets a new medical record.

Hawking's colleagues shake their heads at such dramatic pronouncements. 'Stephen's just Stephen,' a former graduate student of his, Malcolm Perry, now a physicist at Princeton University, said. 'He doesn't take it very seriously, so we don't, either.'

Gerald Wasserburg, a geologist and physicist at the California Institute of Technology, who has met Hawking at a number of conferences, says of him, 'He is one of the most striking examples in the history of science of the power of the human intellect.'

However, Hawking is not without his critics in the tightly knit physics community. One top theoretician at Princeton told me once, 'He's working on the same things everybody else is. He just receives a lot of attention because of his condition.' Other physicists have accused him of being overly dramatic and argumentative at science conferences.

Despite these bouts with temperamental and jealous colleagues, Hawking's work has been widely honoured. In 1978 he received the Albert Einstein Award, considered by some the highest honour in theoretical physics.

In 1982 alone he received honorary degrees from Notre Dame, the University of Chicago, Princeton, and New

York University. The Queen has named him a Commander of the British Empire. The media have frequently labelled Hawking this half-century's answer to Einstein. Hawking himself disputes such notions with one of his standard remarks: 'You shouldn't believe everything you read.'

CHAPTER THREE

Galileo's Eyes

Stephen Hawking suggested to me that Galileo Galilei, the seventeenth-century astronomer, may have been the best scientist of the twentieth century. 'He was the first scientist to actually start using his eyes, both figuratively and physically. And, in that sense, he was responsible for the age of science we now enjoy,' said Hawking.

'And he used his eyes to good effect. He knew what he had seen, and he acted on it. He knew how to draw the right deductions. Once he knew he was right, he stuck with it.' Hawking believes today's scientists, some 340 years after Galileo's death, could use a little more of the same attitude.

'Like Galileo, scientists today have to be prepared to step outside the mainstream, out beyond the currently accepted ideas. That is the way you make progress.' He laughed almost silently for a few seconds. 'Of course, you have to know which way to step.'

Several letters each week come to Hawking from people far out of the mainstream. He is rather amused by them. One he showed me was a savage scrawl of equations on a single sheet sent by a man from Michigan. 'He thinks he may have found the secret to the universe,' Hawking said, 'but this man is no Galileo.'

Galileo was Hawking's — as well as Einstein's and Newton's — direct intellectual forebear in the sense that he was the first to define gravitation, nature's most pervasive yet, paradoxically, its weakest force. Since Galileo, it has been a matter of correcting, redefining and adjusting the original explanation. Newton repaired and refined Galileo; Einstein honed and broadened Newton's basic laws to include the entire universe. Now Hawking and other cosmologists are trying to do the same to Einstein's general relativity, the modern explanation of gravitation and the force that most concerns cosmologists.

In 1905, when he published three papers in Volume 17 of the German scientific journal *Annalen der Physik,* Einstein's ideas were revolutionary; it was then far from apparent that these papers would change the course of science history. The first paper dealt with statistical mechanics, and the second, which he thought the most important, with the photoelectric effect.

The third paper was the bombshell. Destined to change for ever the way we look at time and space, it outlined the special theory of relativity, as it later came to be called, and challenged the old dictum that space consisted of a matter-permeating ether and that time worked like the flow of a river. These were ideas that had dominated science for hundreds of years.

Einstein showed that time and space must be defined in terms usable to scientists – not to poets or philosophers. They had to be quantities that ordinary men using ordinary tools could measure – not scientifically useless abstractions. There was nothing more to either space or time. It was a straightforward, twentieth-century solution to a nineteenth-century problem.

Boldly dismissing the best thinking of the previous two hundred years, Einstein stated two postulates: one was that, no matter the motion of its source, light always travels at a constant speed. This was not news. Every measurement ever taken up to then had borne this out, and it was well established that light travelled about 186,000 miles per second (186,282 miles per second is the precise figure used today). Yet none of the great experimentalists of the day was willing to believe the implications of the evidence that lay right in front of them.

Nobody saw what Einstein saw, that the velocity of light is always the same, that it never changes no matter its source or direction. This held true, Einstein wrote in the third paper, no matter where the light came from. In other words, the speed of light was constant through empty space even if its source was moving very fast – like a galaxy or star.

This was a heretical notion and seemed to violate common sense. It meant that light projected from a star moving towards us would have the same velocity as light from a star moving away from us. It was, and still is, an unsettling thought. It is logical to assume that a bullet fired from a gun aboard a moving train will have a greater speed – the velocity of the bullet in addition to the velocity of the train – than a bullet fired from a gun at rest. The same, Einstein said, did not hold true for light: its speed is always constant, and as a result light's velocity is different from the speed of anything else. A bullet or moon or planet always has a speed that is relative to something else. The speed of light is relative to nothing; it is an absolute constant, always the same.

The other postulate was that an experimenter is able to detect only relative motion. In other words, to a person standing on a station platform as a train speeds by it is the train that is in motion, not the platform. Yet another person on the train could just as well imagine that he and the train are standing still while the person on the platform, and everything else, is flying past him.

These two postulates – one stating that all motion is relative, and the other excepting the speed of light, which is an absolute constant, seem contradictory. Yet in the world of special relativity they do not conflict, and the postulates served to abolish Newton's basic assumption that time is absolute and that, like a river, it always flows from past to present.

To demonstrate the constancy of the speed of light and the relativity of all other motion, Einstein used thought experiments like the following. If a person standing on a station platform sees two lightning bolts, one far to the east and the other far to the west, strike the tracks simultaneously, he would conclude that they occurred at the same time. Yet to a person sitting on a train moving at high speed from east to west just in front of the platform it would look as if the bolt in the west had struck first.

The reason, according to Einstein, was that the observer on the train was moving towards the bolt in the west and, because light speed is constant, its light reached him slightly sooner than from the one in the east. The person on the platform thus saw two simultaneous flashes, while the observer on the train saw first one and then another. They would have reported different phenomena that, in fact, were the same. Moreover, had the bolts struck at slightly different times, the one in the east first, it would have been the person on the train who reported two simultaneous flashes.

Which of the observers was wrong? Both were right, depending on their frame of reference – the train or the platform. By similar reasoning, Einstein showed that time and space were linked and equally fickle, depending on the motion of the observer. Using reasonably simple mathematics he showed, for instance, that to the person on the station platform the windows of the train flying by would actually be shortened. As the train speeded up, approaching light speed, the windows' length would shrink to nothing. To the person on the train, the windows would remain the same.

Nothing remained the same in Einstein's bold new relativistic world – except, of course, the speed of light. Some bizarre conclusions emerged from such thinking. For instance, if the person on the platform could see the watch of the person on the speeding train, the timepiece would move more slowly, even at the slow speeds of normal, earthbound trains. Of course, the slowing down of the watch would be impossible to measure, it would be so slight. But at higher speeds, near the speed of light, the changes would be monumental.

Einstein demonstrated mathematically that to a person on earth watching a spaceship move away at a speed of 160,000 miles per second, about 86 per cent of the velocity of light, a clock aboard the ship would seem to be moving at only half speed. It also would look as if the ship's mass had doubled while its dimensions had shrunk to half their previous size. To an astronaut aboard the spacecraft, the changes seem to be occurring not aboard his spaceship but on earth, where time would also appear as if it were slowing down.

In declaring that time is measured differently for objects or people moving relative to one another, Einstein abolished absolute time for ever (the concept of 'for ever' was another idea that no longer had meaning in the relativistic universe). Einstein later showed that an astronaut aboard the ship travelling close to the speed of light would age more slowly than his twin brother left behind on earth.

In his fourth and final paper in 1905, Einstein made perhaps the boldest stroke of all. He had already abolished previous notions of space and time; now he did the same for mass and energy. Before Einstein, mass and energy were viewed as separate and distinct. Intuition tells us, as it had physicists before Einstein, that a ball and the energy to throw it are not the same thing. Einstein found, from the postulates to special relativity, that this distinction was not valid.

Using mathematics from special relativity and some ideas from his paper on the photoelectric effect, Einstein came to the conclusion that if an object emits energy in the form of light its mass will be reduced by the amount of energy divided by the velocity of light squared – that is,

$m = \dfrac{E}{c^2}$. From there it was just a simple algebraic step to the most famous equation in history, $E = mc^2$, which was published in 1907.

Einstein showed that mass and energy are not merely equivalent, but interchangeable. The implications were enormous. It meant that even a small portion of matter under the right conditions could be converted into an awesome amount of energy, equivalent to the explosive power of thousands of tons of TNT.

That special relativity works, that mass and energy are indeed interchangeable, has been demonstrated thousands of times in particle accelerators, the immense atom smashers physicists use today to explore the nucleus of the atom. At Fermi National Accelerator Laboratory the masses of protons accelerated through a tube four miles in circumference have been found to increase many thousands of times as the velocity becomes a significant fraction of the speed of light.

In working out his special theory of relativity and the postulates, Einstein had dealt only with new laws involving the measurement of space and time between observers moving at uniform velocity, that is, not accelerating or slowing down or travelling along a curve such as a planetary orbit. Einstein knew he had to solve the more complicated questions of accelerated motion.

One of the biggest problems of nonuniform motion involved gravitation, manifested in accelerated velocity as the earth pulls an object towards the ground. The remarkable thing about gravity, noticed by Newton and Galileo, was that it seemed to act the same on all bodies regardless of their weight. In his famous — although probably apocryphal — experiments from the Tower of Pisa, Galileo supposedly had shown that objects of different mass struck the ground at the same instant when dropped simultaneously. If there was any difference, such as when a cannonball hit the ground sooner than a feather, it was due to air resistance.

Galileo and Newton had seen gravitation as a unique force in nature: a force peculiar to the earth or other heavenly bodies. Einstein saw it as a broader phenomenon.

Suppose, he said, a scientist rides in a lift in a spaceship far from the influence of the earth's gravity. Imagine that the lift inside the spacecraft is accelerating upward at the rate of 32 feet per second each second. That is the exact rate an object – like a cannonball dropped from a tower – is pulled towards the earth by gravity. But in the spaceship's lift, away from gravity's influence, the scientist's feet still press against the floor as his body resists its upward acceleration, and if he drops a stone it falls to the floor just as on earth.

The scientist cannot tell whether the downward pull is caused by gravitation or is due to the inertia of his body resisting the upward acceleration of the lift. This means that there is no difference between acceleration caused by gravitation and acceleration from other sources, Einstein said. It was called the principle of equivalence: a gravitational field has a 'relative existence'.

Had Galileo jumped from the Tower of Pisa and then dropped a stone on his way down, both he and the stone would have been in free fall. The stone would have appeared to Galileo to be in a state of rest and, with the effects of gravity momentarily suspended, Galileo could have, for a few seconds, considered himself in a state of rest as well.

Then what *is* gravity? Einstein used ideas from special relativity and added new ones to describe it in a unique way – an explanation that showed gravity wasn't really a force in the usual sense. Einstein added to special relativity a different kind of geometry, since he had found the old kind – Euclidean geometry – too limited for his new way of looking at the universe.

An old friend, Marcel Grossman, whose notes had

helped Einstein pass an important exam when they were classmates in a Swiss high school years before, told him where to look. It was a type of non-Euclidean geometry that had been developed by a German mathematician named Bernhard Riemann. It gave Einstein the mathematical tool he lacked: the geometry of curved space.

But what have curved space and accelerating lifts to do with gravitation? Imagine again, said Einstein, that the lift on the spaceship holding the scientist was accelerated so immensely that its speed began approaching that of light. If that were the case, a beam of light entering through a hole in one wall would appear to the scientist inside to bend a little bit down in an arc and strike the opposite wall at a lower point.

The reason is that, by Einstein's earlier formulation, light and mass are equivalent under certain conditions. Since light has energy, it therefore has a mass, and everything with mass is attracted by gravity. And what is gravity but a form of acceleration? Thus, in the accelerating lift, the light and the scientist are equally affected, and both are drawn towards the lift's floor. By the same reasoning, Einstein said, if a beam of light passes near a heavy object such as a planet, gravitation will actually bend the light path in towards the planet.

Einstein brought these concepts together in ten mathematical formulations or field equations, which he published as his theory of general relativity in 1916. It was even more revolutionary than special relativity, since it had virtually no theoretical antecedents at the time.

The most remarkable thing about general relativity was that Einstein did away with the concept of gravitation as a force. In fact, he said, there was no such thing as the force of gravity. It was instead the geometry of the universe – the curved geometry supplied by Riemann – that was responsible for the force we think of as gravity. Einstein called his curved space a space-time continuum.

It was a bit like a trampoline. If you placed a cannonball on it, a large indentation was the result. An orange would make a smaller dent in the trampoline, and have an inclination to roll towards the deeper hole. Stars and planets have the same effect on space that balls have on a trampoline; heavenly bodies actually put a dent in the space around them, altering the geometry of space itself. Larger objects in this dented, curved space tend, like a cannonball on a trampoline, to pull less massive objects towards them.

General relativity moved far beyond the conventional thinking of the day. It was a new physics altogether, an entirely different way of looking at the universe, and there were a number of disbelievers.

Two natural phenomena existed which Einstein was certain would prove that his ideas about curved space were right. The first involved the orbit of Mercury, which for more than a century had refused to move in the elliptical orbit described by Newton's physics: its orbit had 43 seconds of arc too many when the planet was nearest the sun. Nobody had been able to explain this difference, small but measurable with the technology of the nineteenth century. When Einstein's field equations were applied to Mercury's orbit, they predicted a difference of precisely 43 seconds of arc.

The other test of the theory was more difficult. Einstein's equations showed that light from a distant star would be bent slightly by the gravitational field around the sun – just like the light beam in the scientist's lift in the spaceship. The deflection would be exactly 1.75 seconds of arc, the equations predicted. The only time to test the idea was when the sun was in total eclipse, since the light from any star in line with the sun would be obscured by sunlight.

As it happened, a total eclipse was to occur in the Southern Hemisphere on 29 May 1919, about three years after the publication of the general relativity theory. An expedition was launched by the Royal Society to Principle, an island off the west coast of Africa. And during the eclipse British physicist Arthur Eddington found deflections in starlight that nearly matched Einstein's calculations. When informed of this confirmation in Berlin, Einstein responded that he had never doubted what the results would be. Asked what he would have thought had the measurements not confirmed general relativity, he replied, 'Then I would have felt sorry for the dear Lord.'

With the observations seeming to prove general relativity – the first of many confirmations that the universe behaved in almost exactly the way general relativity dictated it should – modern theoretical cosmology was born.

A reworking of man's conceptual view of the universe nearly always follows a period when the old view has begun to fail. New facts are discovered which do not fit the old scheme of things, and the old conceptual view begins to crumble. Science was ripe for a conceptual overthrow when Einstein appeared on the scene. There were enough chinks in the Newtonian edifice to require a vastly different look at things. Whether there are now enough chinks in the monolith of twentieth-century physics to lead to a new conceptual view is not certain.

Hawking, born into a universe wholly and acceptably described by general relativity, is among the second generation of scientists who grew up with its dogma. Progress in the twentieth century has been so rapid that Einstein has already become less than sacred. Has Einstein's vision of the universe begun to crumble so much that we might be on the threshold of a new age of science? Hawking doesn't quite answer the question. 'One cannot say until it happens. One of the beauties of something undiscovered is that it is undiscovered.'

CHAPTER FOUR

The Einstein Connection

On several occasions Einstein said or wrote, 'God does not play dice with the universe.' This was a declaration of his abiding exasperation with quantum mechanics, the mathematical system developed in the 1920s and 1930s to explain the behaviour of subatomic particles. Decades later Stephen Hawking replied, 'God not only plays dice, but sometimes he throws them where they cannot be seen.' It was not as pithy as Einstein's remark, but it made Hawking's point: time and knowledge have at last overtaken Einstein.*

In his office Hawking has a small collection of photographs and posters of Einstein. Old ones are replaced with new arrivals from time to time. Yet all Hawking would say to me of Einstein was, 'Well, he was a very fine physicist.'

Following experimental confirmation of general relativity, Einstein received much worldwide acclaim. He was received by kings; newspapers and magazines beat a path to his door for interviews; and popular books on general relativity sought to explain its secrets. There was also resistance to relativity. Some people refused to believe that a lone man using only mathematical hieroglyphics could redefine the entire universe.

Despite the acclaim and controversy, Einstein forged ahead with his work. It would have been impossible to

---

* The remark about the dice being where they cannot be seen refers to the possibility that they may be inside a black hole.

duplicate a feat as revolutionary as general relativity, but he wanted to broaden it. Its equations describe the geometry of space-time, and he was certain they would work for the geometry of all space-time – that is, for the universe from its beginning to its end. He published a paper a year later in 1917 that, more than anything else, established modern cosmology – the study of the origin, history and shape of the universe.

It was a remarkable piece. In it he set down the principle of the laser forty years before the first one was made, a stunning achievement in itself. But, more important, he described how the equations of general relativity could describe the behaviour of large pieces of matter in the universe over long periods of time. He ran into trouble right away.

The problem was that the best and simplest interpretations of his equations pointed to an unstable universe, possibly even one that was expanding. Among others, the Dutch astronomer Willem de Sitter had already solved the equations indicating that the universe was nonstatic, either expanding or collapsing, but not standing still. Einstein balked. He wanted his equations to show the heavens as portrayed by most astronomers: stable and unchanging, isotropic – the same in all directions – and homogeneous – the same everywhere.

Einstein found a rather odd way out. In order to make general relativity fit this model of the universe he altered his equations, adding a figure he called the cosmological constant, referring to it as a 'slight modification'.

One immediate problem with the cosmological constant was that general relativity was a theory so complete in itself that it did not need universal constants. So the 'delta terms', as they were called, were really unnecessary. Einstein himself was keenly aware of this, and the last sentence of the 1917 paper declared, 'That [delta] term is necessary only for the purpose of making possible a quasi-static distribution of matter as required by the fact of the small velocities of the stars.'

In 1922 a Russian mathematician, Alexander Friedman, solved Einstein's equations both with and without the cosmological constant. Like Einstein's, his solution with the cosmological constant produced a static universe that remained the same for ever. Friedman's more daring second solution left out the delta terms and led to the first model of an expanding universe, actually two different models. It has yet to be determined which is correct; each posits a different view of the universe's eventual fate.

The two Friedman models of an expanding universe are, in fact, the basis of cosmology today. The first is one in which the density of matter is less than a certain critical amount, meaning that the universe is infinite and will expand for ever. In the second – the one approved by most modern cosmologists – the density is greater than the critical level. The expansion of the universe will, as a result, one day cease. It is finite, but also unbounded; in it, if you start off in a straight line, you eventually come back to where you started.

It's a bizarre concept that we have come to accept as naturally as the notion that eggs fry in a hot pan. Hawking thinks that such a universe, curved back on itself, is like a gigantic black hole that also curves around itself. At least, he says, the mathematical descriptions are similar.

'Another way you might like to look at this model of the universe', says Hawking, 'is as a gigantic expanding balloon. Points on it represent galaxies. As the balloon is blown up, the points move apart from each other.'

In 1922 Einstein published a mathematical criticism of Friedman's work. He soon withdrew it, and dropped the subject altogether for nearly a decade.

About the same time, a succession of larger and larger telescopes being built in the western United States were seeing in the heavens what Friedman had already predicted in his calculations. The 60-inch reflector was built at Mount Wilson in California in 1908, and in 1917 a 100-inch telescope was built at the same observatory. Edwin Hubble, a former amateur boxing champion, began working at Mount Wilson in 1919, and by 1923 he had made the first estimate of the distance between our galaxy, the Milky Way, and the Andromeda galaxy, our nearest neighbour.

Hubble also showed that Andromeda was about the same size as the Milky Way, the first hint that other parts of the universe were the same as ours. During the 1920s Hubble discovered that distant galaxies are distributed evenly across the heavens and, more important, confirmed that they are all flying apart from one another like spray from a shotgun.

Hubble announced in 1929 that his data showed that the galaxies were moving apart at a rate directly proportional to their distance from the Milky Way. This was the first direct proof that the universe was expanding; it became known as Hubble's Law and convinced most physicists that Friedman's interpretation of Einstein was more correct than Einstein's interpretation of Einstein. Einstein came to admit that the cosmological constant was the worst mistake in his scientific career.

Before Hubble's observations, Friedman's calculations had served as little more than a plaything for theoreticians. But linked with Hubble's Law they established what is known today as the cosmological principle. Its basic tenet is that the universe is the same, roughly, in all directions. The universe will also look approximately the same to any observer no matter where he finds himself stationed in the cosmos. Since the 1930s, virtually every observation has confirmed expansion of the universe, but whether the cosmological principle holds is not necessarily borne out by astronomical data.

'In fact,' Hawking said to me as we talked about this

proposition, 'there is no guarantee that the universe is the same at all points. One is eventually led to a picture in which the universe could have different branches. We could be in a branch of the universe that doesn't allow us to see all the rest of the universe. In fact, there is a non-zero probability for the universe to have many different forms.'

Still, general relativity coupled with Friedman's interpretation and Hubble's observations provided for the first time a complete – if not necessarily entirely accurate – picture of the universe. Yet it was only when Hawking and Roger Penrose appeared on the scene decades later that cosmologists began to realize how comprehensive that picture actually was.

'One of the features of the Friedman solution that was not taken too seriously at the time was that this solution indicated there was a singular epoch in the past in which all the matter of the universe was concentrated into a single point,' Hawking told me. This is the point known as 'singularity'. 'Most people felt at the time that conditions in the real universe could never have been so extreme.'

Hawking told me that, at the time when he and Penrose had put their minds to the problem of a relativistic interpretation of the universe's origin, the Friedman model had offered a reasonably good view of what had happened back to the first one hundred seconds or so. 'Naturally, we were rather eager to find out what happened before that,' he said.

The problem with the Friedman model, as innovative as it was, was that the real universe contains irregularities. 'As one went back in time, these might grow large and cause the individual particles converging to miss each other, giving a sort of nonsingular bounce,' Hawking said. 'In that case, the points would have missed one another during the contraction and then the universe would actually have re-expanded without ever reaching the singularity.

'As a result, nobody took the Friedman model very seriously as an interpretation of what occurred during the creation of the universe,' Hawking told me. 'In fact, most people thought that there was no true beginning. We proved them wrong.'

By running Friedman's model of the expanding universe backwards in a theoretical contraction – in a sense, going backwards in time – physicists were searching to discover what happened at the very beginning of the universe. More fundamentally, they wanted to prove that the universe had, as Friedman's model suggested, a beginning in which all matter was concentrated in a single point, and that there was a Big Bang in which this point emerged and exploded, creating our universe, space and time. Hawking and colleague Roger Penrose were to accomplish this feat.

Penrose, a young mathematician and theoretical physicist, was then at Birkbeck College at the University of London. He had already established himself as one of the world's foremost mathematicians. He was a master of geometrical and mathematical riddles and puzzles, and had provided the inspiration for several drawings by the Dutch graphic artist M. C. Escher. Penrose's interests were no accident – his father was a noted geneticist and inventor of mathematical puzzles and one of his brothers a ten-time British chess champion; his uncle was a leading surrealist painter and a friend and biographer of Pablo Picasso.

'The first major area we worked on was whether or not time had a beginning or will end,' Hawking told me. 'At the time that I started working on the problem in 1962, the general opinion was that it did not have a beginning.'

One logical place for Big Bang theoreticians to look for observable models is in the phenomenon of collapsing stars. These stars, collapsing in on themselves of their own weight, may eventually lead to a black hole at the core of which exists the problematic 'singularity'. Most important, they have some of the characteristics – in reverse – of an expanding universe.

The history of any star – whether of average size like the sun or as big as Antares, whose diameter is as large as the earth's orbit – is essentially a tug-of-war between the powerful outward-directed force of its heat and radiation – the product of interactions taking place within the star's atoms – and the strong inward-directed force of gravity. If a star were heavy enough, none of the other three interactions at work in the universe – the strong nuclear force, the weak nuclear force, and electromagnetism – would be able to resist gravitation's pull on the star's own matter. It would begin collapsing in on itself.

What is to prevent such collapse from continuing for ever, the star crushing itself down to an infinitesimal speck containing all its matter, a single point of infinite density? Imagining an endlessly collapsing star, physicists were unable to determine what would happen when a star reached the point they called singularity. Singularity is the end of the road, a place where space and time simply disappear. 'At singularity, normal concepts of space and time break down,' Hawking told me. 'The same thing happened to the equations.'

Many theorists had believed that singularities would turn out to be nothing more than mathematical abstractions. It took a brilliant mathematical *tour de force* from Penrose to show that an endlessly collapsing star was not simply a theoretical plaything, and that it would end as a real, a physical, singularity. Penrose showed that space and time can come to a physical, rather than a merely metaphorical, end.

In 1965, Hawking, drawn by this proof of singularity, started working in collaboration with Penrose. During the next three years they developed several key theories about the structure of space and time and singularities, showing that it was with a singularity that the universe began.

Demonstrating that the universe began as a singularity of infinite density, similar to the end product of a star in ultimate collapse, was no easy task. 'That, of course, was the point where all of our equations broke down,' Hawking recalled. 'There were solutions to Einstein's equations that most people in those days thought were not realistic; they represented a universe that was too uniform and isotropic.

'Most people working on the problem believed that to get close to the truth you had to have a complicated solution with numerous irregularities on a large scale,' Hawking said with a laugh. 'Nobody wanted to believe that the truth could be as simple as it was.'

As the expansion of the universe is played backwards in a theoretical contraction, one of the problems physicists face is the possibility that the particles, with irregular and random motion, will hit each other. This was the thinking of a group of Russian theorists in 1963, when they proposed a theory that called for alternating contracting and expanding phases during the Big Bang that would allow particles to avoid striking each other.

Hawking laughed. 'My first big piece of research was to show they were wrong.' Between 1965 and 1968 he worked on the problem. Hawking explained their thinking to me. 'We developed a new mathematical technique that was actually an analysis of the way that points in space-time can be causally related to each other. The thing is that in general relativity no signal can travel faster than light. So two events cannot be related unless they can be joined and you can get from one to the other at a speed equal to or less than that of light.' This meant that Hawking and Penrose did not need to explain what would happen to individual particles at the instant of the Big Bang, as everyone else had been trying to do. 'In fact, we found that you could use large-scale properties to prove that there must have been a singularity at the beginning, a much simpler approach. This meant that time has a beginning.'

Hawking and Penrose went on to prove that not only could the universe begin in a singularity, but that it *had* to begin in a singularity.

'What we did was show that the simplest solution of general relativity was the correct one,' Hawking said. 'In fact, given the overall complexity of the universe, it actually is quite remarkable that this is so.'

Friedman's interpretation of Einstein's general relativity had produced the first picture of a complete universe. Hawking and Penrose added to his rough draft a relativistic interpretation of the early universe that required the existence of at least one physical singularity.

This singularity that Hawking and Penrose uncovered with their calculations, although certainly real and physical, is mathematically an event in space-time where normal physical behaviour breaks down. Since their calculations showed that a relativistic universe possessed of real physical traits must have such a singularity, the implication was clear: you cannot look at the universe using general relativity without finding a Big Bang or something very similar at the beginning. The theoretical analysis of the Big Bang – the first ever – was one of the biggest steps in cosmology since Friedman had applied his mathematics to general relativity.

About the same time – during the 1960s – astronomical discoveries confirmed Friedman's theoretical work. The most important of these (after Hubble) was the background radiation distributed evenly throughout the universe and found unwittingly by Arnold Penzias and Robert Wilson in 1964. 'It was correctly interpreted as a relic of the Big Bang,' said Hawking. 'It had been predicted by George Gamow and his colleagues back in 1948, but at the time, partly as a result of the problems with the

Friedman model, nobody took the prediction very seriously.'

Another finding of huge significance was the discovery that the element helium comprises about 25 per cent of the mass of all matter in the cosmos, the other 75 per cent being made up mostly of hydrogen. 'Gamow's calculations and later refinements had predicted that about one hundred seconds after singularity one quarter of all the protons and neutrons that had been created earlier should have changed into helium along with a small amount of deuterium. It was difficult to account for so much helium in any other way than in the way Gamow's calculations had shown. So it was a satisfactory finding for theorists, nearly as much as the discovery of the background radiation left over from the dense phase of the universe.'

These discoveries confirmed the work of theorists showing that the universe had its origin in an intensely hot explosion. Doing so, they made clear that theorists were capable of more than mere speculation, that they could come to some valid conclusions about the birth and history of the universe. And Hawking and Penrose brought all these observations together, explaining how the Big Bang was not just factually and theoretically plausible, but even necessary.

Ironically, Einstein, in spite of the great error of the cosmological constant, had reached a similar conclusion years before. When he dropped – with Friedman's indirect help – the egregious delta constants, he found that general relativity required that the universe be populated with at least one singularity during its history. But he dismissed this singularity as a point in the equations where theory simply broke down.

According to the most accepted version of the Big Bang today, all the material in the universe comprised an extremely hot compressed gas in a primordial fireball 10 to 15 billion years ago.

'The biggest misunderstanding about the Big Bang is that it began as a lump of matter somewhere in the void of space,' Hawking said as we continued our talk on the very early universe. 'It was not just matter that was created during the Big Bang. It was space and time that were created. So, in the sense that time has a beginning, space also has a beginning.'

Does he actually believe that time, in fact, did begin with the Big Bang? I asked him. 'When you get far back into the very early universe, the ordinary concept of time becomes obscured. You cannot have the normal idea of time carried back indefinitely. There is some point near the Big Bang when there is simply no way to define time. In that sense time has a beginning.'

The physics of general relativity was used in order to prove the existence of the singularity at the beginning of the universe. 'The problem with this approach is that general relativity, which was used to predict the singularity at the origin, is a purely classical theory,' said Hawking. 'So there is nothing in general relativity to take into account the quantum behaviour of subatomic particles that were created in the Big Bang.'

The movement and mass of subatomic particles is described by quantum mechanics. Quantum mechanics, the mathematical system developed during the 1920s and 1930s, is wholly alien to general relativity. It describes the interactions that take place at the subatomic level, and at its core is the uncertainty principle first announced in 1927 by the German physicist Werner Heisenberg.

The uncertainty principle states that certain pairs of quantities, such as the position and momentum of an electron, cannot be measured simultaneously. This means that the electron is not the objective, absolute and determinable bit of matter that classical physics describes, but a sort of objective entity that in a sense is smeared out around the nucleus.

The uncertainty principle distinguishes quantum mechanics from all other physics because it declares mathematically that atomic and nuclear particles are distributed in an uncertain and random fashion. The location at any instant of any particle can be described only by using a system of probabilities and statistics. Einstein insisted on viewing the universe as an orderly, predictable place. General relativity was a perfect reflection of that view. To Einstein, the quantum system was philosophically and mathematically unequipped to exist in the same universe as general relativity. Today's physicists, though, consider it of equal importance to general relativity. And, like general relativity, quantum mechanics has met every experimental test ever devised for it. These experiments are conducted in particle accelerators that break apart the constituents of atoms to find out what they are made of, a process that some theorists caustically liken to smashing a watch to see what falls out.

Quantum mechanics seems to suggest that the subatomic world – and even the world beyond the atom – has no independent structure at all until it is defined by the human intellect. (This view of the universe has similarities to Eastern philosophy, which has led, to the dismay of Hawking, to a wealth of popular literature such as Fritjof Capra's *The Tao of Physics* and Michael Talbot's *Mysticism and the New Physics* that attempts to link quantum physics with Eastern mysticism.) Physicists have been unable to reconcile this system with the view of the universe posited by general relativity. While general relativity allows for a perfect pointlike singularity at the beginning of time, quantum mechanics does not, for it prohibits defining at the same time the precise location, velocity and size of any single particle or singularity.

Ultimately, quantum mechanics will have to be brought into play if we are to understand the workings of the infinitesimal universe at its very beginning. Only by reconciling the two seemingly irreconcilable areas of physics can theorists hope to find the unified field theory that will explain the workings of the entire universe.

Hawking's work has shown that the formulation of such a theory will also take a deep understanding of black holes, which in their bleak and forbidding structure contain key mathematical similarities to the beginning of time.

CHAPTER FIVE

—•—

Black-Hole Encounter

The publication of a cover story on black holes by *Time* magazine on 4 September 1978 was the culmination of popular fascination with these unseen and often misunderstood objects. Hawking was mentioned prominently in a side note to the article, and referred to as 'one of the premier scientific theorists of the century, perhaps an equal of Einstein'.

Hawking laughed when I asked him about the comparison. 'It's never valid to compare two different people – much less two different physicists,' he said. Closing the subject, he added, 'People are not quantifiable.'

Hawking doesn't dispute the notion that he is a master of black holes, and although his attention is now focused elsewhere – mainly on the very early universe – he still looks on black holes with awe and amusement. He is always ready to talk about them.

'You can't get there from here,' he said with a grin when I asked him what an encounter with one would be like. Then, with a barely detectable shrug of his slight shoulders, he asked how deeply I wanted to go into the mathematics as he started explaining in detail his work with black holes.

Black holes, as Hawking tells it, are rips in the fabric of space and time so dense and distorted by unimaginable gravitational forces that for years physicists believed nothing could escape from one, including light. They are thus, by definition, invisible. No one has seen or ever will see one, no matter how powerful the telescope.

Hawking is certain they exist. 'There may be as many as

a thousand million in our galaxy alone,' he said. I asked him for the evidence, and he conceded that for the moment their existence can be confirmed only as special solutions of the equations of general relativity and by a few scanty parcels of indirect physical data.

Despite the mystery, physicists in recent years have begun calling on black holes, largely as a result of Hawking's work, to explain everything from the creation of galaxies and quasars to the ultimate fate of the universe itself.

'It's a little like using the unexplainable to explain the unexplainable,' Hawking told me. As much as anybody else, he delights in the enigma of these most mysterious of all celestial bodies. 'Within black holes, space and time as we normally perceive them come to an end. It's a disturbing thought.'

An object like an asteroid or astronaut coming too near the edge of a black hole would first be stretched out of shape like a rubber band – then vanish into the hole without a trace. In that sense, black holes are cosmic vacuum cleaners that suck up everything they encounter, from giant stars to particles of space dust and the photons that make up light. There is no escape from a black hole.

Hawking and other theorists are convinced that the long-sought unifying concept in physics – the theory that would explain the central interaction of the universe – lies at the periphery of black holes or similar peculiar constructions arising at one point or another in the evolution of the cosmos.

Such a mathematical construction should, at least in theory, be able to explain the construction of every bit of matter in the universe as well as all the forces that interact between this matter – somewhat akin to concocting a single recipe that would work for soup and cement and everything in between, all expressed mathematically. Far-fetched as it sounds, Hawking assures me that physics is within twenty years or less of such an all-encompassing concept.

When I asked Hawking how he first became interested in black holes, he told me, 'It was in black holes that it first became apparent to me that the strong forces that bind the elementary particles could come together with the weaker forces of gravitation. And of course black holes have a certain appeal all of their own, in their mystery and the images they convey to the human mind.'

Black holes are the natural consequences of the death of stars. If the collapse of a star may lead ultimately to a singularity, a black hole might be described as the final stage in the death of a star before the point of singularity is reached. And it is the black hole that finally masks the singularity from the rest of the universe, creating a break with the ordinary space-time around it.

Using only Newtonian ideas about gravity and light, the French scientist Pierre Simon Laplace first suggested in 1796 what might happen if a star were large enough. He theorized that there could be a strong enough gravitational attraction to recapture all the star's radiation, including light.

'It is possible that the largest luminous bodies in the universe may actually be invisible,' wrote Laplace, nearly two centuries ahead of his time.

In the case of the sun, there has been a standoff between the contending forces for close to 5 billion years. Astronomers think it will continue in equilibrium for at least that much longer. At that time, it is theorized, gravity will at last begin winning the tug-of-war as the sun's nuclear fuel is spent. The mass of the sun, a ball of dense, hot gas 865,000 miles across, will then begin to collapse.

When the sun's matter has become condensed enough, it will become what astronomers call a 'white dwarf', a figurative name that refers to a seething ball of atomic nuclei and loose electrons that, in the case of the sun, would be only about four times as large as the earth, minuscule in cosmic terms.

Its mass would remain about the same as it is now, however. As a result, the gravitational pull on the atomic matter at its surface would be far stronger than it is today. The velocity required for an object like a rocketship to escape from the sun's surface would have increased from 380 miles per second today to over 2100 miles per second.

The collapse can continue. Physicists are certain that a star can collapse indefinitely. To get to that point of permanent self-annihilation, the star has to be massive indeed. In the case of the sun, with only average initial mass, it will collapse no further once it becomes a white dwarf. A law of physics called the exclusion principle intervenes at that point.

This law states that two electrons cannot occupy the same energy space, meaning there is a limit to how tightly matter can be packed together. This limit is high by ordinary standards: at the white dwarf stage, a thimbleful of the sun would weigh tons.

If the original star has a greater mass — estimated to be 3.6 times or more that of the sun — the exclusion principle will be overpowered by gravitation. Such a star going through the throes of gravitational collapse would be drawn down further, breaking atomic nuclei apart, destroying atoms.

It eventually becomes a 'neutron star', a heavy mass of neutrons just a few miles across. Escape velocity at the surface would be 120,000 miles per second. If the star is more than about three to six times the mass of the sun, it will not stop contracting at the neutron star stage. Gravity is clearly in charge now, and it shows no mercy. It draws the star down into itself, a victim of its own weight. Finally it reaches the point where the escape velocity at the surface is so great it reaches 186,282 miles per second — the speed of light. If you watched a star at that exact instant, its already dim glow, not much more than a weak electromagnetic shadow, would flicker out.

Gravity claims light as its final victim. The former star, now a black hole, has become absolutely invisible and will remain so for a very, very long time.

Astronomers have been able to pick up traces of white dwarfs that still radiate enough light to be photographed by large telescopes, and the electromagnetic squawks of neutron stars can be detected by radio telescopes. By their nature, however, black holes are poor correspondents. There is general agreement that they exist, but astrophysicists – and even theoretical physicists – would dearly like to have a look.

As for Hawking himself, does he really believe that black holes exist beyond the figment of an equation? In fact, he has become convinced, along with some other physicists, that at least one has been found.

'If you look at the constellation Cygnus, there is a good chance that you will be looking in the direction of a black hole,' he maintains.

Some stars travel in pairs. They are called binaries and orbit a common centre of gravity. Astronomers reason that if one of the stars in a binary system has collapsed into a black hole the invisible black star would still hold its gravitational embrace on its visible mate. Hawking is certain astronomers have found one such mixed marriage in the constellation Cygnus, 6000 light-years from earth. 'The visible star, a blue one, is stretched and distorted,' he says. The reason: its mate, a black hole, is exerting tremendous gravitational force on it, pulling it into the shape of an egg.

The discovery in 1973 of this apparent black hole in the binary system called Cygnus X-1 has excited theoretical astrophysicists far more than if another planet had suddenly drifted into sight beyond Neptune (which now, by a temporary quirk in orbit, is the most distant planet from the sun). Its origins are the subject of endless speculation.

Hawking has made a bet with one of his best friends, Kip Thorne, a respected theorist at the California Institute of Technology, about the ancestry of the mysterious object in Cygnus X-1. If it turns out that the binary system does not contain a black hole – breaking many a physicist's heart – Hawking will win a four-year subscription to *Private Eye*. If it is a black hole, Thorne wins a one-year subscription to *Penthouse*.

This odd four-to-one bet has become notorious in physics circles. Why should Hawking, whose work virtually requires the existence of black holes, bet against them? 'It really is a statement about my own psychology,' he told me one day as we talked about the probability that Cygnus X-1 contains a black hole. 'Actually I could win easier than Kip. Any number of observations – such as the emission of pulses – could disprove a black hole.'

However, he is certain that Cygnus X-1 will turn out to be the real thing. 'If it isn't a black hole, it really has to be something exotic,' he said.

Astronomers may have found more than one. A team of Canadians and Americans announced in 1983 that they had discovered a second black hole, this one outside our own galaxy. They found it, by its emission of powerful X-rays, in the Large Magellanic Cloud, a satellite galaxy of the Milky Way visible only in the Southern Hemisphere.

Using the 158-inch telescope at the Inter-American Observatory at Cerro Tololo, Chile, they estimated the black hole's distance from earth at 180,000 light-years, its weight about ten times that of the sun, and its distance from its binary mate a mere 11 million miles.

Within Cygnus or the Large Magellanic Cloud or elsewhere, a black hole by any definition is a weird resident of the universe. Its existence strains the laws of physics. Moreover, what is to prevent a black hole from collapsing even more – in fact, down to a singularity, an infinitesimal speck of infinite density like the one at the onset of the Big Bang?

Hawking and Penrose demonstrated in their early work that this is precisely what could be expected to happen in the case of some burned-out stars. Later, Hawking, working with other colleagues, was able to show that a black hole would probably settle down into a fairly stationary state that was no longer related to the star from which it had collapsed. In fact, such black holes would be possessed of only three measurable parameters: mass, rate of rotation, and electrical charge.

'This turned out to be of actual practical importance,' Hawking said. I had asked him what difference it made if an object that could not be seen or measured did have actual parameters.

'Well,' he explained, 'it meant, for one thing, that the structure of the gravitational field of any black hole could be accurately predicted. And that meant that one could construct models of astrophysical objects – such as Cygnus X-1 – thought to contain black holes. The properties of the model could then be compared with actual observations.'

In the 1970s black holes began making an appearance as a cultural phenomenon. They were a common topic on chat shows, and the subject of endless jokes. The public passion of the mid-1970s for black holes was perhaps just a fad. In that sense black holes were a sort of Bermuda Triangle of space, something lying somewhere between parapsychology, the occult, UFOs and astrology.

The image of a black hole or the Big Bang both confuses and delights our subconscious. Black holes might be a metaphor for our own fate or for the fate of the universe. If a star can crash down in upon itself, why not the entire universe?

'Calling these things black holes was a masterstroke' said Hawking of John Wheeler, the American physicist who christened them. 'The name conjures up a lot of human neuroses. There is undoubtedly a psychological connection between the naming of black holes and their popularization.

'It is important to have a good name for a concept,' he said, speculating for a moment on the psychology of scientific terms. 'It means that people's attention will be focused on it. I suppose that the name "black hole" does have a rather dramatic overtone, but it also is very descriptive. It has a strong psychological impact. It could be a good image for human fears of the universe.'

Just as there was utter oblivion before the Big Bang, there is utter oblivion at the centre of a black hole. Normal time ceases to exist as surely as it did not exist before the Big Bang. Therein lies much of the appeal of black holes and the Big Bang for Hawking. Moreover, in both concepts come together the twin pillars of twentieth-century physics: Einstein's theory of general relativity and Max Planck's quantum theory. In 1974 Stephen Hawking, in a bold and risky theoretical gambit, proposed a startling new idea about black holes that hinted for the first time that quarks and quasars actually might operate within the confines of a single, although deeply hidden, law of physics.

CHAPTER SIX

— ◆ —

Exploding Black Holes

The Rutherford–Appleton Laboratory is a few miles off the M4 motorway, about an hour and a half's drive west of London in the dry and unscenic flatlands south of Oxford. Built across several square miles, Rutherford–Appleton is the British equivalent of the Los Alamos National Laboratory near Santa Fe, New Mexico.

Although scientists at both places do basic research in particle physics, theoretical physics, and energy, these laboratories are primarily workshops for designing nuclear weapons. Despite diverse geographical locations, the two labs have certain similarities: insecure-looking wire fences, indifferent middle-aged guards, and relatively few checkpoints, all belying the highly classified work going on inside while creating an ambience of studied casualness of the sort preferred by the physicists working there.

In the winter of 1974 Hawking took a trip down from Cambridge to Rutherford-Appleton. His purpose was to deliver a paper. It was one over which he had agonized for months. He was still agonizing over it the day he was to present it to a group of fellow-physicists. The paper was called 'Black Hole Explosions?'

Usually cocksure, Hawking was concerned about how the paper would be received. What he was proposing was a radically new idea that, if right, would force a basic rethinking of theoretical physics. The question mark in the title was a reflection of his own doubts. It was as if Isaac Newton had been sufficiently uncertain to issue a pamphlet called 'Does Gravity Draw Objects Downwards?'

Despite Hawking's credentials, the audience was stacked against him. The lecture hall was filled with particle physicists and accelerator experimentalists. They can be the bane of cosmological theorists, whose work runs too much to mathematical formulation and too little to practical results to suit most lab physicists. A stray nuclear-weapons physicist or two had also dropped in.

Hawking began. The lights were dimmed, and slides of his equations began flashing on a screen. As Hawking continued, it became apparent that he had come to a startling conclusion about black holes, an idea utterly contrary to the conventional wisdom of the day.

Most of the physicists there had trouble following Hawking's argument. There were few questions, and Hawking quickly concluded his presentation. As the lights were snapped on, the moderator, John Taylor, a professor of mathematics at the University of London and the writer of such popular books as *Black Holes* and *The Shape of Minds to Come* (a speculation into parapsychology), stood up and declared, 'Sorry, Stephen, but this is absolute rubbish.'

The paper Hawking presented that day at Rutherford–Appleton had its origins in work he had first undertaken in 1970. Interest in black holes had been greatly spurred two years before, in 1968, with the discovery of rapidly pulsing cosmic radio sources. After some initial confusion among astrophysicists these were generally interpreted as being caused by rapidly rotating neutron stars, objects with a mass about the same as the sun, but with a radius of no more than ten miles.

They were being called 'pulsars', and seemed to confirm the existence of neutron stars, collapsed stars composed almost wholly of neutrons packed as densely as in an atomic nucleus and thus so heavy that a cupful would weigh tons. A neutron star is not a black hole, merely a way station *en route* to a black hole, a resting point in the collapse that occurs when a star's gravity overpowers the outward thrust of its nuclear furnace.

This apparent confirmation that neutron stars exist showed that theories about the progression followed by a star when it collapses were essentially correct. It was an easy step in 1968 for cosmologists to say, 'If neutron stars exist, then why not black holes, too?' Unlike a neutron star, a black hole by definition could not give off any sort of radiation. Its only apparent effect would be its gravitational influence on a nearby star. Evidence of this came in 1972 with the discovery of Cygnus X-1, the binary system that prompted Hawking to make his bet with Kip Thorne.

Three years before, in 1969, Roger Penrose invented a thought experiment suggesting that a black hole could exert more than simple gravitational influence on nearby matter. He proposed that energy could be extracted from a black hole – if it were rotating. The idea was called 'superradiance' and suggested that certain types of wave in the vicinity of a black hole would be amplified and thrown off – rather than absorbed – by the rotating black hole.

Penrose's thought experiment also hinted that some of the rotational energy of the black hole itself could be carried away. This was the first attempt to show that a black hole did not have to exist as an entity unto itself, sundered from all other matter in the cosmos. A rotating black hole could lose electrical or rotational energy through a process known as pair creation. The idea was that a particle and its antiparticle – an electron and an antielectron or positron, for instance – are formed just outside the hole.

Then, say the electron is drawn by gravity into the black hole, but the positron escapes. In the process, a tiny portion of the black hole's own electrical charge is cancelled and a minute fraction of its angular – or spinning –

momentum is carried away. The black hole has actually lost energy, something never thought possible before.

In those days Hawking was thinking about the boundary around a black hole, the precise point at which light can just escape the powerful grasp of the black hole's gravitation. This is called the 'event horizon'. The more massive the black hole, the larger the surface area of the event horizon.

The event horizon can be envisioned as a sort of one-way membrane through which light can penetrate from the outside, but can never exit from the inside. An observer sitting inside the black hole could see flashes of light entering the hole, a coded message from a spaceship just beyond the event horizon.

The observer, though, would not be able to send a signal back. The light or the radio wave or any other form of energy would get no farther than the event horizon. The captain of a spacecraft waiting for a message from a scout who has gone into a black hole to see what it was like and report back would wait for ever.

The event horizon, which foils the captain and his scout, is somewhat less frustrating to the theoretical physicist. In fact, for Hawking and other earthbound physicists it holds some fascinating implications. One of them is this idea that once light – or anything else – falls into a black hole it becomes invisible to an outside observer. Physicists have expanded this notion of lost information to what they call the 'black holes have no hair' theorem.

This odd expression, typical of the sort physicists like, means simply that two black holes with identical mass, electrical charge, and spin look the same to an outside observer regardless of what they are made from. Even a black hole made of matter and another of antimatter would be indistinguishable, meaning that most of the physical features of a black hole are forever invisible to an observer.

To account for these unseen features, physicists realized that the size of a black hole's surface area – the size of its event horizon, in other words – was its only significant feature as far as somebody on the outside was concerned. This was the single feature that could be explained in terms of an actual, meaningful number, since everything else about the black hole was hidden from view.

The particular significance of the size of the event horizon, its most interesting aspect, was on Hawking's mind towards the end of 1970. One night as he went to bed an idea came to him, one so obvious that he had trouble sleeping the rest of the night. It was simply this: the size of a black hole's event horizon, its surface area, can never decrease. It is a straightforward concept that anybody can understand without the aid of mathematics.

'Well, one must get an idea someplace,' Hawking said later of the idea's birth. The next few days after that sleepless night, he and a few colleagues tested the idea mathematically. It seemed to hold up.

Hawking had applied the ideas of general relativity to reach his conclusion about a black hole's event horizon. Think about the spaceship captain sending his scout for a look at the inside of a black hole. The scout, having left the spaceship in his little rover craft, would approach the hole directly, then drop through the event horizon. This is the way it would appear to the scout, assuming he did not drop into oblivion.

To the captain on the bridge, it would appear that the scout approaches the black hole in a slow – and endless – spiral around it, the apparent speed of the scout's craft directly related to the spinning velocity of the black hole itself, if it were spinning. The captain would never actually see his scout penetrate the event horizon.

The different time-scales are explained by an exaggerated case of time dilation, the same thing experienced by a space voyager travelling at the speed of light. Near the event horizon the black hole's gravitation is so great it pulls all objects – including the advance scout – at an accelerating velocity approaching the speed of light.

To the captain watching from his spaceship safely out of the pull of the gravitational field, the scout appears to travel exponentially slowly, just as the space traveller flying along at light speed appears to an earthbound observer to age not at all. The rate at which the scout's rover craft appears to the captain to be slowing down is thus inversely proportional to the pull of gravity at the event horizon. The more massive the black hole – and the greater the gravitation – the slower the scout's apparent deceleration to the captain. The opposite is true for the scout: the bigger the black hole, the slower he thinks he is moving through the event horizon.

To Hawking, general relativity's idea of time dilation meant that the size of the event horizon itself could never appear to an outside observer to shrink. The idea was a significant step in the theoretical investigation of black holes, because for the first time it set a universal limitation on the behaviour of all black holes: an event horizon cannot decrease, only increase. Prior to this insight there had been no such thing as a static or dynamic boundary for a black hole.

Hawking's idea of an unshrinkable event horizon also established an important link with the concept of entropy, which by definition also increases with time. The notion of entropy, a corollary to the second law of thermodynamics, declares that the amount of energy available to perform a physical task must always decrease. *Entropy* is the word that defines this gradual 'inutility' of energy as it is transformed from one type to another, say, from electricity to heat.

A form of energy that is highly useful like electricity has low entropy, and energy like heat with lower utility has high entropy. Energy with low entropy can always be

Stephen Hawking in his office in Cambridge.

A Hawking family portrait: from left to right, Lucy, Stephen, Robert Timothy, and Jane.

Hawking in his study at home with Timothy and Robert.

Hawking with students at Cambridge University.

The Large Magellanic Cloud, a satellite galaxy of the milky way, emits
powerful X-rays which scientists believe may be evidence of a black hole within

S-17, a gaseous cloud, which may also contain a black hole.

Fermi National Accelerator Laboratory. Located in Batavia, Illinois, Fermilab is one of the largest particle accelerators in the world. Using these 'atom smashers' scientists are discovering a new and varied world of subatomic particles. And these particles may ultimately have something to tell us about how the very early universe worked and whether a unified theory can be found to explain all phenomena.

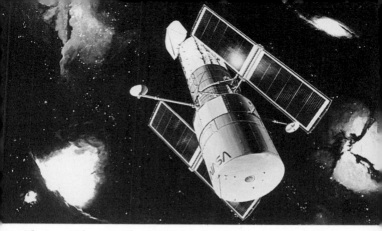

The Space Telescope. When NASA launches the space telescope in the late 1980s it will enable scientists to look further into space than ever before. Physicists predict that what they see there will confirm the work of theorists like Hawking concerning the existence of black holes and singularities.

Roger Penrose, Hawking's collaborator. Together in the 1960s they successfully proved that time does indeed have a beginning.

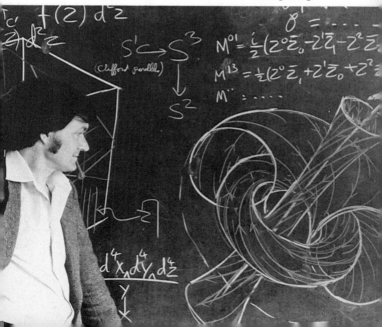

Hawking receiving the Lucasian professorship of Mathematics at the University of Cambridge.

converted to energy with high entropy. So it is easy to convert electricity to heat. To reverse the process is impossible. Entropy can never decrease. It always increases in the sense that the inutility of energy increases. This holds true for any system – an automobile, a computer, a star, or the cosmos. Since the size of the event horizon of a black hole also never decreases, it was reasonable to assign any black hole an entropy value that would describe many of its physical characteristics that otherwise were permanently invisible to outside observers.

This, however, led to a major difficulty: if an entropy value were assigned to a black hole, then a temperature also had to be given it in order to maintain a workable thermodynamic model. But an object with temperature must radiate heat, and black holes, by their classical definition, emitted nothing at all. It was a problem that Hawking was not to resolve for several years – until he wrote the paper that he presented at the Rutherford–Appleton Laboratory.

In the meantime he did not give up working on black holes. In 1971 he postulated that the death of a star was not the only way one could come into existence, and that the universe could contain millions of black holes that were not downgraded stars at all.

Until this time theorists thought that the only gravitational field powerful enough to form a black hole was that of a collapsing star of ten or fifteen solar masses. But what would cause a small black hole – one the size of the nucleus of an atom, for instance? Could gravity draw any kind of material, stellar or otherwise, down to that size? This was difficult to fathom, since gravity has almost no effect inside something as small as an atom's nucleus.

Hawking, though, arrived at an unexpected conclusion: if sufficient pressure were applied to any object, even one only a kilogram or two in mass, it could be compressed to enormous density. At some point, when the material of a small object like a chunk of metal has been squeezed into a small enough space — say, the size of a proton — self-gravitation would take over. The compression would continue until a tiny black hole appeared.

But a force large enough to initiate such a process exists nowhere on earth, or for that matter in the visible universe. The answer, Hawking knew, lies at the beginning of time. Only then was the material from which the universe formed located in the same place at the same time and under sufficient pressure to produce small black holes.

In a perfectly uniform universe this could never occur, Hawking knew. But if there were irregularities, such as regions far denser than others, the areas with a certain excess density could have collapsed into little black holes. He calculated that these mini-black holes could have been squeezed into existence in overdense regions of the rich primordial soup that existed within the unimaginably small fragment of time occurring in the first 10^{-20} second after the Big Bang.

It is possible, he now thinks, that the universe is populated with billions of these tiny but potent black dots. He expects that they could be as close together as the distance between the earth and Pluto. This translates to 100 million mini-black holes per cubic light-year throughout the universe.

After the publication of Hawking's paper, physicists embraced the idea of these mini-black holes. They offered answers to several previously unexplained phenomena, and possibly supplied the so-called missing mass that astro-theorists think may be hidden somewhere in the universe. News of the idea quickly swept through the worldwide community of physicists.

Hawking continued his work on small black holes, taking his next and most daring step: positing that something as tiny as one of these primordial objects might, like an atomic particle, consent to being described by quantum mechanics.

In 1973 quantum mechanics and general relativity were

universally thought to be incompatible. Yet Hawking thought it was time to explore the possibility that black holes, themselves a key aspect of general relativity, could be expressed in quantum terms.

Later that year Hawking started thinking about the behaviour of matter in the vicinity of black holes, large or small. By autumn he had the glimmer of an idea. One day as he thought through the mathematical Sanskrit of black holes, he made a discovery so radically opposed to previous ideas that he was sure he had made a serious mistake.

Hawking had found that black holes, in defiance of every known law of physics, emit a steady stream of particles. But like everybody else he believed the theory that black holes could not emit anything, unless (possibly) when rotating. He spent weeks afterwards trying to find the flaw in his calculations.

What had convinced him that black holes could emit particles was the application of quantum theory at the black hole's edge – precisely at the event horizon. His reasoning was that the uncertainty principle, the backbone of quantum mechanics, declares that empty space is never really empty. It is always active and cluttered. Pairs of elementary particles like electrons and their antimatter opposites, positrons, exist for a fraction of a second before uniting and annihilating each other in a little burst of X-rays.

If such a transaction were carried out at the event horizon, one of the particles could be caught in the powerful embrace of gravity and wander down into the black hole, never to be seen or heard from again. Instead of joining in mutual annihilation, the other particle would then be free to escape outwards. To an observer it would look as if this second particle had just popped out of the black hole. (Any larger object, governed by the rules of general relativity, would be prohibited from participating in a quantum transaction of this sort.)

In late 1973 this was a bizarre and unexpected notion. Hawking still disbelieved his own results. He spent weeks working on the problem. At last he was sure he had solved the problem that had been holding him up: Where did the energy for the quantum transaction at the event horizon come from? The answer, which Hawking had suspected all along and was able finally to demonstrate mathematically, was that the black hole itself, with its enormous gravitational field, supplied it.

Hawking's calculations thus produced another stunning conclusion: with its energy going into the production of particle emission − or radiation − the black hole itself would slowly erode over time. Eventually, when its gravitational field had become so dissipated that it could no longer hold itself together, the hole would explode, bursting apart in a shower of high-energy gamma rays.

Not all black holes would explode. In the case of large black holes such as those created by the collapse of a star or the monstrous holes that astronomers think may exist at the cores of some galaxies, the evaporating process would take too long − many times the age of the universe. In the case of a black hole with the mass of the sun, for instance, or another average-size star, Hawking figured that the evaporation and eventual explosion would not occur for 10^{66} years after that sun had become a black hole. Even in cosmic terms that is a period of time so large as to be meaningless.

His calculations showed, however, that small black holes could evaporate and explode far more quickly − their average life expectancy being about 10 billion years, a long time but somewhat less than the age of the universe. He figured their average radius to be 10^{-13} centimetres, about the size of a proton, and they would weigh about a billion tons more than a proton, or approximately the weight of Mount Everest.

The emission of particles at the event horizon would be

enormous. Hawking calculated it at around six thousand megawatts, roughly the same as the output of six large nuclear power plants. However, utility companies should not begin planning to use a small black hole as a power source; if one were brought to the earth's surface, its weight would propel it through the planet like a bullet through a down pillow.

The final stage of one of these small holes' evaporation would go so fast that it would cause a tremendous explosion equal to the blast from 10 million one-megaton nuclear bombs. These were not idle calculations, and Hawking was certain they had real meaning in the real universe. In fact, he was able to show how his theory about exploding black holes could be proved or disproved experimentally, a crucial element of any new scientific theory.

Since the lifespan of the mini-black holes was close to the age of the universe, he figured that some of them should be exploding from time to time just about now. These explosions would produce a massive outpouring of high-energy gamma rays. Hawking thinks it would be an easy matter to detect these outbursts with a large gamma-ray collector in orbit around the earth. He has suggested that the space shuttle could construct one, and estimates that there are perhaps two black-hole explosions per cubic light-year per century occurring in our corner of the universe.

Still, when he first arrived at it, the idea of evaporating and exploding black holes was so unorthodox that even Hawking questioned his own results. He sat on his findings for weeks, going over and over the calculations in his head. It was too weird, too bizarre. Nobody would believe it. In effect, he was saying, 'A black hole is not a black hole when it erodes and explodes.' It was too much even for Hawking.

In his book *The Structure of Scientific Revolutions*, Thomas Kuhn, the eminent historian of science, points out that significant scientific breakthroughs are rarely accepted when they are first announced. A scientist who makes a breakthrough – a Copernicus, a Galileo, or a William Harvey – is likely to be ignored or even ostracized for years.

In early 1974 Hawking feared this would happen to him. He knew that if his ideas about exploding black holes were right it would revolutionize astrophysics. The idea of solid black holes was firmly entrenched. If Hawking were proved wrong, it could take him years to regain his credibility.

So he waited, going over and over the calculations in his head and talking only to a few close friends and colleagues about the black-hole emission that wouldn't go away. It did not help that some of his colleagues questioned the results. One day Martin Rees came up to Dennis Sciama and asked, 'Have you heard? Stephen's changed everything.'

During those days and weeks of uncertainty, Sciama quietly urged Hawking to announce the results. Sciama at last prevailed, and Hawking decided to make the trip to Rutherford–Appleton. The initial reaction proved worse than he had feared. Could he actually be wrong? Hawking wondered briefly.

The next month he published his results in *Nature*, the prestigious science weekly. Within days physicists around the world were discussing it. In the weeks that followed, exploding black holes came to be among the most talked-about new ideas at physics conferences. A few physicists even called the new theory one of the most significant findings in theoretical physics in years. And Sciama, a Hawking fan all along, pronounced the paper 'one of the most beautiful in the history of physics'.

The black-hole dynamics Hawking uncovered had many important implications. They were similar – in reverse – to some of those thought to have occurred during the instant following the Big Bang. So it looked as if this model could help physicists understand the way in which subatomic particles were created and interacted with one another during the moments of explosive genesis. More important, by applying quantum mechanics to black holes, Hawking had taken the tentative first step towards finding the underlying interaction that might bring quantum mechanics together with general relativity. Their unification – the so-called quantizing of gravity – is the most difficult problem in physics.

Since 1974, mathematical evidence that black holes do emit particles and explode has accumulated and been confirmed by other theoreticians using different approaches. The black-hole emission itself is called 'Hawking radiation', and any physicist anywhere can tell you exactly what it means.

CHAPTER SEVEN

─────── ◆ ───────

The Final Question

There is nothing harder to envision in the universe than a black hole. Except the Big Bang, which has, nevertheless, ignited many visions in the minds of physicists. And in these mind, the Big Bang has been occurring over and over – in countless calculations in thousands of heads – for only a bit more than a quarter of a century.

A nonphysicist might see something like this: into a void, so absolute as to mock any human concept of emptiness, appeared a single point of raw potential. And at the very instant of its creation this point, bearing all matter, all dimension, all energy and all time, burst out, spewing forth its contents.

At the instant of its origin, all matter and all forces were indistinguishable from each other.· As the universe expanded and cooled, matter and force split apart and then split again. Still in the first billionth of a second in its history, the universe continued to fragment. Soon all the constituents of matter – what we now call quarks and leptons* – assumed separate identities, falling into separate classes that have never been joined again.

The single force propelling the cataclysm also became fragmented, with new forces splitting away as quarks and leptons were formed, and the different particles became associated for ever with the new forces that were being

_____

* Leptons are the class of subatomic particles which includes the electron, neutrino, tau and muon – all objects that exist outside the nucleus which is composed of neutrons and protons, themselves consisting of quarks.

created. Three of the fragmented forces are still at work inside the atom. The most powerful of these is the strong force that holds the constituents of the nucleus together — the quarks that make up protons and neutrons. One thousand times weaker is the electromagnetic force that keeps electrons, a type of lepton, in orbit about the nucleus. This force makes atoms appear solid and also is responsible for radio and light waves.

Another hundred times less powerful is the weak force that causes radioactive decay by slowly breaking down neutrons in atoms of certain elements like uranium. All the forces are transmitted by vector bosons — force-carrying particles that exist for a fraction of a second, transmitting force in much the same way energy would be transmitted between people in rowing boats throwing a ball back and forth between them, recoiling at each toss.

Vector bosons live just a fraction of an instant while they impart their force. It is as if the ball tossed between the boats disappeared after each catch. A vector boson called a gluon is responsible for the strong nuclear force, and the photon, a massless particle that outside the atom is the constituent of light, is the boson responsible for electromagnetism.

In the case of electromagnetism and the strong nuclear force, the bosons behave like balls flying back and forth between jugglers, the transactions holding the jugglers together while exchanging energy between them. Three particles — two called W's and one called a Z — transmit the force responsible for radioactive decay. They were first discovered by a team headed by Carlo Rubbia at CERN (a French acronym for European Council for Nuclear Research, located in Geneva) in 1983.

Gravitation, the weakest and only other known force at work in the universe, is some 10^{38} times less powerful than the strong nuclear force. An as yet undetected vector boson called the graviton is theorized to cause gravitation, a force that has no meaningful effect inside an atom because it operates only on large mass.

Only during the past decade have the grand unified theories emerged attempting to show that the three sub-atomic forces are components of a single underlying inter-action. None yet includes gravitation.

Sheldon Glashow, co-winner of a Nobel Prize for his work with grand unified theories, once told me, 'When the universe was very, very hot, we believe that all the forces may have been one. And that one underlying, seemingly magical force is what we are all now working to discover.'

Hawking agrees. 'To unify the four forces in a single mathematical explanation is the greatest quest in all science.'

I asked him how he proposed to make this quest, and what was his own personal goal.

'My goal is simple,' he said, suddenly solemn. 'It is a complete understanding of the universe, why it is as it is and why it exists at all.'

I scribbled these thoughts into a notebook. When I looked up, Hawking was convulsed with laughter and there was a gleam in his eye.

'Do those words sound familiar to you?' he asked.

After an instant's reflection, they did indeed. A year earlier I had written a story about Hawking that had appeared in a popular American science magazine. In a prominent position early in the story, Hawking was quoted uttering those exact words, which he had spoken to me when I had met him the previous year.

Still, Hawking, like most theoretical physicists, now believes that the secret of the most elusive of all goals lies in the very early universe, the period within the first trillionth of a second after the beginning of the Big Bang. It was then when the four forces we see in our cold, stable universe were probably one. This has been suggested by examining events at the incredible instant when the universe was but 10^{-32} or 10^{-33} second old. But, Hawking thinks things need to be pushed back even further than that — by more than a factor of a billion, in fact.

'I would like to know exactly what happened between 10^{-33} second and 10^{-43} second,' Hawking said. 'It is there that the ultimate answer to all questions about the universe — life itself included — lies.

'The classical notion of time would break down somewhere before then — somewhere between 10^{-33} and 10^{-43} second. I recently wrote a paper that was mainly concerned with the universe at 10^{-33} second, but my real concern is what lies beyond that. I don't think there is a very definite model at this point. Nothing seems to provide the answer so far.'

There are a number of constraints upon any explanation of the beginning of time. First is that nobody was there to observe it. Any theory purporting to describe the Big Bang is an enormous extrapolation from the evidence that still exists today. The radiation background throughout the universe is one of mankind's major links to the creation. Like coals found in a circle of rocks in the woods, a strong clue that something hot was there before it shows physicists what the universe was like when matter and radiation split apart from each other. Because it is so uniform and so omnipresent, it indicates that cosmological models of a homogeneous — or isotropic — universe are essentially correct. The other essential piece of evidence that cosmologists must contend with is the 75:25 per cent hydrogen–helium ratio in the universe today. Hawking and most other physicists think this was produced when the universe was but a few minutes old.

I asked Hawking how scientists could be certain that they were at last beginning to fathom the earliest moments of the universe. Was it not possible that cosmologists might be overlooking entire eras in the development of the universe or seriously misinterpreting basic observations?

'Well, it is possible,' he said. 'But remember that we can always look backwards in time by looking out farther into space with our telescopes. The farther into space we look, the closer to the beginning we come.'

'But doesn't that mean making an assumption that everything out there and everything here is the same and operates the same way?' I asked.

'It does.'

'And doesn't it suggest that we believe that the natural laws we have discovered in our time have always been at work in the universe?'

'It does indeed.'

'But are not scientists in making those assumptions taking a leap of faith that is more metaphysical than scientific?'

'One makes some rather strong assumptions in delving into the beginning of the universe, but most of the facts – such as the background radiation – seem to bear them out. And there is no reason so far to believe that our calculations are incorrect in any essential way.'

In fact, observations in space and inside powerful accelerators have drawn a remarkably consistent scenario on the trip back through time. On this journey there are several key stopping-off places, points of special significance where theorists like to examine events in detail.

The first of these is when the universe was a billion years old. It is then that astrophysicists believe quasars – now thought to be the most distant objects in the heavens – began to form. How they formed is still one of the central topics of cosmology conferences today. But it is generally agreed that at about this time the universe began to take on its familiar look, bright spots in a black sky.

The next stopping-off place is over 10 billion years ago, when the universe was a mere 500,000 years old. It was then that elementary particles joined together to form atoms. Before then the universe was still too hot for an electron to fall into a quantum orbit around a nucleus, and the cosmos was a boiling sea of loose electrons and nuclei.

Physicists can mark this point because they know exactly how much electromagnetic force is required to keep an electron attached to the nucleus of any atom of an element. It is only necessary to convert this force into a temperature equivalent and see at what point the cooling universe passed through that stage. The answer is half a million years.

According to Hawking, 'Once atoms had begun to form, then matter could condense into galaxies and stars, and gravity could start playing an important role in the development of the universe.' It is also at about this point that light, had there been anybody to see it, began to be able to travel throughout the universe.

The next stopping place on the journey backwards is at about 100,000 years. The two chief ingredients in the universe are the matter that makes up the galaxies, stars, planets and people, and the radiation that constitutes the microwave background. Today the radiation background and matter have almost no interaction with one another.

Yet in the universe's early age – when density and temperature were far greater than now – matter and radiation interacted strongly. Theorists believe that the photons of the microwave background, in fact, were coupled with the protons, neutrons and electrons that make up matter. As the universe cooled, radiation and matter parted company, this occurring after 100,000 years. The microwave background discovered by Penzias and Wilson in 1964 is a relic of the instant radiation split off from matter.

The next stop back for theorists is in the infancy of the universe – at about three minutes past time zero. 'This is an important point,' said Hawking. 'Before three minutes, the universe was too hot to allow protons and neutrons to join in a nucleus. It is at three minutes that we have to start looking at the strong force very carefully.'

If a proton and neutron did happen to come together during the first three minutes, collisions with photons from the background radiation or with other particles would kick them apart. At three minutes, however, things had cooled sufficiently for the strong force to begin pulling a proton and a neutron or a proton and two neutrons together to form a nucleus of heavy hydrogen.

At about the same time, nuclei of helium were being created out of a pair of protons and one or two neutrons, and it was then that the 75:25 hydrogen–helium ratio that exists to this day was established. Nuclei of a few other light elements were also formed at the three-minute mark, but it was not until millions of years later that heavy elements like iron and gold began to be forged in the furnaces of the stars.

The picture is quite clear at three minutes. The interactions of the strong force, now dominant, are well understood. Even medium-size accelerators can create conditions quite similar to those existing at this time.

Continuing the countback through the first three minutes, the universe is mainly in a cooling-down phase. At about one hundredth of a second after time zero, the universe is hot enough – about 200 billion degrees Celsius – for hundreds of types of particle to be created and annihilated in the energy produced in collisions with one another. This hundredth-of-a-second milestone is important in another sense. Physicists are fairly sure that they have got the facts right in their backward journey to this point.

'The scenario is clear,' said Hawking. 'There is little disagreement up to a hundredth of a second.'

From this point backwards in their search for the lost history of the universe, cosmologists have been forced to rely on the work of particle physicists, whose accelerators are still fully capable of creating similar conditions on earth. In other words, even moderately powerful accelerators can still match the universe's energy at this point.

Between one ten-thousandth (10^{-4}) and one millionth (10^{-6}) of the first second is another milestone. Then the basic constituents of matter in the universe were created as quarks combining in groups of three, forming neutrons and protons. Before this, the universe was a seething soup of quarks, too energized and too densely packed to form into nucleons.

Physicists actually use the notion of a 'quark soup' to overcome certain theoretical problems they face at this point. At 10^{-4} second the universe's density is so great that the amount of space between all protons and neutrons is as small as the size of one of the particles themselves. That is a little too close for the comfort of physicists who have been forced to write a new description of the way subatomic particles behave if they are to account for such density.

Fortunately for them, one of the properties of the strong force, which holds the nucleons together inside an atom's nucleus, is that this force increases with distance. The closer the particles, the less the force. This is true for protons and neutrons and for the quarks that make them up.

So physicists have developed models in which the high-density early cosmos is a mix of barely separated particles that do not interact with one another – the so-called quark soup model. It proposes that the temperature simply gets higher and higher towards time zero. If the model is correct, it could mean that the temperature keeps building right on through the Planck wall, the point at which all equations simply break down.

Other models project what is called a cold Big Bang. This does not mean a universe that began in a cold explosion but, rather, an expansion of space that might not have been hotter at the beginning than a trillionth of a second later. In models of this sort, quarks are not necessarily involved. Instead, proponents suggest that the number of elementary particles increases without any apparent limit towards the beginning of the universe.

This seems like a paradox. The universe historically has always cooled down. It is cooling down today. So as we look farther and farther backwards in time the universe should be hotter and hotter, right up to the instant of the singularity at the beginning. These models, however, claim that the earliest universe's substantial energy went into the creation of an ever-increasing number of more and more massive particles, so at the very beginning we see a relatively cool Big Bang.

Among theoreticians, the quark soup model currently enjoys a greater vogue than the cold Big Bang, although as Hawking says with a twinkle, 'Theorists' minds can change – and quite often do.'

Within the first 10^{-10} second after its birth, the universe had already grown to about the size of our solar system, a colossal expansion of energy. Still, the largest accelerators on earth, in collisions involving a few protons and antiprotons, can duplicate the heat or energy level of the universe when it was one millionth of a second old.

Physicists encounter one of the most significant points of all at this time. It is here that theoreticians first begin to see the unification of the four forces that hold sway in our universe today. Before this time, according to a theory worked out by Glashow, Steven Weinberg and Abdus Salam, the electromagnetic force that controls leptons and the weak force that causes radioactive decay were one and the same.

'If these calculations are right, then before 10^{-10} there were only three forces in the universe,' said Hawking. 'The combined electromagnetic and weak force, the strong nuclear force and gravity. It begins to suggest that there was a common origin for everything we see today in the universe.'

The unified theory of Glashow, Weinberg and Salam can be tested in the very largest accelerators on earth. In 1982 and 1983 a team headed by Carlo Rubbia used the large proton–antiproton collider at CERN near Geneva, to produce W and Z particles with virtually the precise properties predicted by the unified theory.

I visited CERN at the time Rubbia's team was turning up the first W particles. The energy level produced in the accelerator – currently the most powerful on earth – was over 100 billion electron volts, equal to a temperature of 1000 trillion (10^{15}) degrees. This is roughly the same as the temperature of the universe when it was 10^{-12} second old.

The excitement was palpable and contagious in the control room at CERN as hints of the heavy particles started showing up on computer terminals. 'We have been on the track of the little beasts for years,' Rubbia told me. 'They are starting to show up in our carefully laid trap. If they are what we think they are, it is a milestone for experimental physics.'

As they press on in the journey backwards theoreticians are on their own from this point on. Although large accelerators are being constructed, nothing currently being built on earth can ever come close to matching the temperatures of the very early universe, and it will no longer be possible for experiments to verify theoretical speculation.

At 10^{-20} second we come to the place at which Hawking thinks the small black holes may have been formed. In these mini-holes, which Hawking believes now populate the entire universe, physicists must for the first time begin to look at quantum effects in terms of gravity.

'There can't be too many of these,' said Hawking. 'Otherwise, we would observe many gamma rays – and we don't. In our galaxy they can't be any closer to each other than the distance from the earth to Pluto. The gravitational effect of the galaxy would mean, moreover, that the density of the little black holes would be higher inside than outside the galaxy.'

At 10^{-32} second the universe is only about the size of a softball. Its temperature is 10^{27} degrees Kelvin. A microflash earlier – at 10^{-35} second – is the last stop on the trail back where physicists have any confidence that their ideas about the history of the very early universe are correct. The fraction of a second since singularity is a *trillion* times shorter than the amount of time it takes light to cross a proton.

Physicists with faith in the grand unified theories believe that at 10^{-35} second the strong force was one with the other two subatomic forces – the electromagnetic and weak forces. These grand unified theories attempt to describe what the universe was like during this instant in cosmic history. The universe is but 10^{-24} centimetre in size, and its pure energy is just beginning to condense into pointlike particles such as quarks and leptons. Matter and antimatter are present in nearly equal amounts.

The wall of Planck time comes at 10^{-43} second. At this point there is a fundamental breakdown in the ability of physicists to describe space, time or matter. It is presumed that gravity has just broken its bond with the single unified force that existed at the instant of the Big Bang. But nobody knows for sure because there is no quantum treatment of gravity.

To cross the Planck wall, it would be helpful for physicists to know whether the various GUTs are really on track. There is, however, no certain way to test them in accelerators on earth. The energy levels at this time in cosmic history are too high ever to be duplicated, so new experimental techniques may have to be developed.

'If the GUTs are right, then all that is left out is gravity,' said Hawking. 'But I'm not fully convinced that GUTs are the right model.

'And, as you know,' he said, 'there are some very serious problems with the standard Big Bang model. And the various GUTs don't paint a necessarily consistent picture.'

For one thing, there is the persistent problem of the magnetic monopole, an elusive particle most versions of the GUTs predict should have been twisted out of space in the instant just this side of the Planck wall. The theories that explain the creation and dominance of matter have also resulted in these monopoles. They are free magnetic poles, similar to the north or south pole of a magnet existing on its own.

Blas Cabrera, a Stanford University physicist, claims to have found evidence that these monopoles – an altogether new type of matter – actually exist today. But the jury is still out on Cabrera's work.

Another concern is that the microwave background radiation is too uniform. If the radiation were actually released when it separated from matter at a cosmic age of around 100,000 years, how is this homogeneity throughout wholly different parts of the universe explained? At that point in the universe's history, pieces of the sky would already have separated by millions of light-years of space. There would have been no possibility for an exchange of energy over such distance. So why has the same background temperature been detected everywhere?

And still there is the wall at 10^{-43}. There is no immediate hope for getting past this last frontier, where on the far side may exist a self-explanatory universe of ultimate simplicity.

'At that point the gravitational field has become so strong that quantum effects must – by definition – be taken into account,' said Hawking. 'If we want to understand how the universe began, we must understand how gravitation and quantum mechanics are combined.

'The Planck time represents a breakdown in our ability to describe space and time in classical relativity theory. It is because we do not know how to quantize gravitation.'

In a sense, theorists have been able to approach but not cross the Planck wall because they have been able to fudge their equations this way or that to carry them backwards over crucial points of time in the universe's development. This happened at 10^{-4} second, when it was decided it was all right to skip over the interactions between densely packed protons and neutrons.

The Planck wall is the universe's ultimatum: there will be no further hedging of equations, no more jumping over points too complex or obscure to understand. This is where all your calculations and all your thinking must be brought together in the clearest statement in the history of mankind about the cosmos before you will know how it began. And still you may never know – exactly.

CHAPTER EIGHT

—◆—

Bubble or Bang?

The Big Bang has been the *modus operandi* for the creation of the universe for a generation of physicists. Yet visit a cosmology conference today and you will hear nothing but talk of bubbles and the inflationary universe.

Theorists have never been very comfortable with some of the fudge factors in the Big Bang scenario of a cataclysmic explosion in which matter and forces were born and burst out in all directions to form a universe that is still expanding today. The singularity at the beginning bothers some, and for most the Big Bang does not adequately explain the uniformity of the background radiation. It is as if, they think, a bomb exploded, throwing shards out in a perfect sphere. The Big Bang also makes an assumption about the nature of space in the universe, simply presupposing that it resembles a flat tabletop rather than the surface of a sphere, to explain the universe's own flatness.

Hawking points out that 'the idea that the universe began as a bubble offers a simple solution to many of these problems'. The bubble theory, originated in 1981, eliminates most of the flaws in the Big Bang theory while, naturally, creating other problems. The basic idea is that not one but several universes arose from bubbles inflating like balloons in the void. At the very beginning there was an uneven mix of energized spots, each hotter than a trillion quadrillion degrees. Because of their heat, these spots expanded so rapidly that they soon lost their heat. Then they 'supercooled'.

This occurs sometime after the Planck barrier at

10^{-43} second, when gravity already has broken free from the three quantum forces. The supercooled regions have a special property that Harvard physicist Sidney Coleman calls a 'false vacuum', which allows them to draw energy from the gravitational field that surrounds them. At about 10^{-35} second after the process begins, the super cooled regions get an enormous kick of energy, perhaps akin to a jolt of negative gravity, and erupt into bubbles that are destined each to become a universe unto itself.

The 'false vacuum' is the key to the concept. It has been likened to water that has been superchilled well below the freezing point. The water can exist in that state as a liquid for just an instant before crystallizing quickly into ice. The same holds true for the supercooled regions. They exist for an instant in an undisturbed state.

As the bubbles created from the false vacuum's kick begin to cool again, the combined electromagnetic and weak forces split from each other and take on separate identities. The energy inside one of the bubbles – our universe – begins to condense into particles like leptons and quarks. At the end of this inflationary epoch – at 10^{-32} second from the very beginning – our universe, now containing the matter for all the stars, galaxies, planets and people that inhabit it today, is about the size of a grapefruit. At this point the ordinary Big Bang scenario takes over.

Physicists are currently very intrigued by the bubble scenario. For one thing, during the universe's inflationary era, the bubbles developed more slowly than in the Big Bang format. This allowed the matter developing inside time to mix evenly, come to the same temperature, and thus produce a uniform radiation throughout the universe.

The bubble theory also accounts for the flatness of our universe by relying on natural law rather than arbitrary assumption. As a bubble grew to the size of today's universe, it flattened out in the way that the surface of a large sphere like the earth appears flat.

What are the problems still to be worked out in the bubble scenario? Astrophysicists are not exactly sure how galaxies and stars congealed out of clumps of matter inside the bubble that became our universe. And some physicists are troubled by the notion of a multitude of universes created in the same instant.

Is that something we will ever know for sure? Physicists disagree even on this point. Common sense would indicate that, like bubbles in a rising loaf of bread, separate universes will never meet, but some maverick theorists believe with characteristic fervour that, as expansion slows down and stops, some of the individual universes could eventually converge.

'Inflation accounts for the fact that the expansion of the universe was fine-tuned,' Hawking explained. 'For one thing, this allowed the universe to expand as it has without collapsing back on itself like a black hole. On the other hand, matter could also have been spread too thin for galaxies to form.'

I wondered about the problem of the universe at singularity, the 'beginning of time', in Hawking's language. Does the bubble concept help eliminate the singularity that so troubles many theoretical physicists?

'Well, certainly you might be able to construct a cosmological model without a singularity using a bubble,' he said. 'But I don't really think it would help the singularity in a gravitational collapse. And it still would not get rid of the singularity in black holes.

'In short, I don't think the bubble concept would get rid of the initial singularity. But it is possible,' he said, uncharacteristically uncertain.

Several physicists have been at work on the bubble universe. One, a young Princeton astrophysicist named J. Richard Gott, has proposed that our universe is but one of perhaps an infinite number that were created like bubbles in a hot liquid of great but finite density. His calculations have indicated that each of his universes is itself 'open' or infinite in terms of its expansion, and will continue to grow for ever.

These bubbles form out of a type of space named after Willem de Sitter, who first studied it in 1917. It is a specific solution to Einstein's general relativity field equations that describes an infinite curved universe that is always expanding. To his solution, de Sitter added a cosmological constant similar to a universal repulsive force, a sort of negative gravity. In the expansion of the universe as defined in de Sitter terms, any individual point in the universe tends to move away from another point at a constantly increasing rate.

In Gott's version of the bubble universe, the initial bubble forms smoothly out of the surrounding de Sitter space. In the process, the singularity at the beginning is eliminated. According to Gott, the large-scale uniformity of the universe could only have come about if every part of the universe in the early period of expansion were directly, or causally, related to every other part, even if just for an instant.

In standard Big Bang cosmology, parts of the very early universe were causally disconnected because they were too far apart for light to travel between them as the colossal expansion began. This has always troubled cosmologists. In Gott's model, as in inflationary universe cosmology, there is a brief phase of constant density that allows enough time for all parts of the universe to relate causally with each other, thus smoothing out the rough spots and creating a homogeneous cosmos.

One peculiar aspect of Gott's cosmology, however, is its application of Hawking radiation to the early universe. Gott, looking at the interplay between gravitation and quantum mechanics, determined that the event horizons surrounding black holes continuously generate thermal radiation. He has come up with the idea that Hawking radiation accounts for the background radiation scattered evenly throughout the universe.

Hawking and others have already shown that black-hole radiation at the event horizon is just a special case of a profound theorem: that anywhere there is an event horizon thermal radiation is emitted. This means that at the boundary of any region from which light cannot escape — the edge of a black hole or the frontier of a universe – there will be an emission of some type of thermal radiation.

Gott claims that one of the significant properties of de Sitter space is that it is filled with event horizons and Hawking radiation. It is the ever-increasing – in fact, exponentially increasing – expansion that creates all those event horizons. If two points start out sufficiently far apart and separate too rapidly to allow a light beam from one to reach the other, an event horizon will occur between them.

Hawking and his former student and office mate, Gary Gibbons, had already calculated the Hawking radiation associated with these particular event horizons. But Gott takes it a step further.

Using Hawking's and Gibbons' result, which describes mathematically the energy density of the radiation in terms of the expansion of the early universe, Gott adds an additional factor to make that energy density consistent and homogeneous. This constant is more or less the same, in mathematical terms, as filling de Sitter space with a fluid of uniform density. However, this Gott fluid has a negative pressure, what its creator describes as universal suction.

Gott maintains that recent calculations in quantum field theory have produced results showing that Hawking radiation would behave in exactly this odd fashion under certain conditions, specifically those conditions present during the expansion of de Sitter space in its early stage.

His calculations have an interesting result. The event

horizons generate Hawking radiation. The radiation becomes the fluid that causes the bubble universe to expand. Gott's constant – or the fluid of the Hawking radiation – is what causes the exponential expansion of the de Sitter space. And the expansion generates event horizons.

This circular thesis works only if the Hawking radiation is enormously hot – over 10^{31} degrees Celsius – and incredibly dense, an incomprehensible 10^{93} grams of matter per cubic centimetre. Gott is also fairly sure that these extremes of density and heat are just about the right conditions under which gravitation begins behaving suspiciously like a quantum field – the point at which gravity is quantized.

Gott has included an inflationary epoch in his bubble theory. In this phase transition, matter subtly changes. Gott suggests, for instance, that previously massless quarks might suddenly take on mass. During this period, Hawking radiation enters the bubble from the expanding de Sitter space and in a flash – 10^{-42} second – changes into ordinary matter. It is this nearly instantaneous conversion of radiation to matter that Gott believes astrophysicists today look back on and call the Big Bang.

Hawking radiation generated by the event horizons in the new universe is thus responsible for all matter and energy in the cosmos today. Because Hawking radiation is naturally uniform in de Sitter space, the microwave background and the universe itself are both homogeneous.

The risky part of Gott's model is that it tries to tell us what happened on the other side of Planck time – at 10^{-44} second or earlier. During this period, he suggests, other universes – an infinity of them – also could have formed, just like bubbles foaming in a head of beer.

Unfortunately, though, we'll never be able to observe one of these other universes. Each is separated from the other by an event horizon, the light barrier that prevents the transmission of all information from one to the other.

The essence of any scientific theory is that it must be confirmable. The corollary to that is that it also must be disconfirmable. So how does Gott suggest that his speculation about bubbles emerging in a froth and an infinite number of universes be proved or disproved?

More work is needed to determine more precisely the behaviour of bubbles in de Sitter space, for one thing. That would give his model a better theoretical framework. On the factual side, Gott maintains that more complete and better astronomical observations must be taken about the large-scale characteristics and behaviour of the universe.

Most astrophysicists think that galaxies and clusters could never have formed in a totally homogeneous universe. But Gott's theory calls for just such a universe in the instant following its birth. Inconsistencies and random fluctuations had to develop somewhere during the universe's history. Gott believes that a better look deeper into the sky will explain this anomaly.

I asked Hawking about Gott's theory. What did he think of the use of his radiation to account for all the matter and all the energy in the universe?

'In de Sitter space, you have thermal radiation, and it is important for bubbles. But Gott did not exactly take it into account,' said Hawking. He hastened to add in defence of his own work: 'Gary Gibbons and I first discovered that there is thermal radiation in de Sitter space. The reason it occurs is that there are event horizons, just as in black holes. So it is very similar to black-hole radiation and thus similar to Hawking radiation.'

What, then, does Hawking think of Gott's attempts to apply Hawking radiation to the early universe?

'Well,' he said, with his patented mischievous grin, 'I think that Gott – not God – has received an undue amount of publicity. A number of other people have suggested similar ideas, some of them long before him. Other people also have given greater detail of the mechanics that produced the bubbles.

'Two of them are Alan Guth of MIT and Starobinskii in Moscow,' said Hawking. 'In fact, Starobinskii was really the first one to come up with the concept of bubble universes.'

Hawking joined the bubble–bang fray in late 1981 when, attracted by what he considered excellent work by several Soviet theorists, he journeyed to Russia to find out what they were learning about the inflationary universe. There he visited A. A. Starobinskii, A. D. Linde and others at the Lebedev Physics Institute in Moscow.

'Some of the Russians' versions of the bubble universe concept were really very attractive,' Hawking said. 'The basic idea I was interested in was fairly simple.

'The thing is that if you form bubbles, then you are liable to come up with more than one bubble. And these bubbles are likely to collide. And this will give rise to an inhomogeneous universe. And that is not consistent with what we observe today.'

Linde suggested in a paper that he thought it was possible for a single bubble to form without an adjacent bubble forming alongside. This idea was the main thing Hawking wanted to investigate. He spent hours discussing it with Linde in Moscow. 'I left thinking Linde's version was the best of the Russians', but I still realized there was something wrong with his work,' said Hawking.

When he returned to Cambridge, the first thing he did was sit down with his colleague Ian Moss and hammer out a paper aimed at repairing the flaws in the Russian theories.

'The problem with Guth's scenario was that it led to a very inhomogeneous universe dominated by a few large bubbles emerging from the inflationary stage,' said Hawking. 'We showed in our paper that, in certain circumstances, you can have an inflationary period that occurs simultaneously at all points of space in the very early universe. Thus, it creates no inhomogeneities.'

Hawking, who thought up the approach, and Moss, who did the calculations, arrived at this solution by a relatively straightforward process. 'The other papers had treated the exit from the inflationary epoch as a problem in flat space-time,' said Hawking. 'They neglected the curvature and finite horizon of the universe.

'We showed that the inflationary era does not occur in flat space-time but, rather, in curved space-time,' he said. 'The result is that a universe without the inhomogeneities of the other exists from the inflationary epoch.' With its rather simple solution to a thorny problem, the paper drew widespread interest among cosmologists.

In June 1982 Hawking and Gary Gibbons hosted a conference at Cambridge on the very early universe – the first second of its life. Many leaders in the field came: five cosmologists from the Soviet Union including Linde and Starobinskii, along with Guth and twenty-four others from the United States and Europe. One problem drew the most interest.

There was a fatal flaw in all the inflationary scenarios. It was that the universe, although homogeneous on a large scale, is not quite uniform on a smaller scale; it contains clumps of matter in the form of galaxies and stars and clusters of galaxies. In other words, it was simply not clear that an inflationary universe could produce the stars and galaxies we observe in our universe today. The physicists at the three-week workshop broke off into separate groups – headed by, among others, Hawking, Guth and Starobinskii – to solve this problem.

In testing and probing what they had started calling the 'new inflationary universe', the assembled theorists gathered in small clusters around blackboards and computer terminals and thought for a while that they had come up with a solution to the galaxy formation problem. Their calculations showed that the inflationary scenario did indeed produce the right number of clumps of matter distributed properly throughout the universe.

But if the inflationary scenario were followed to its mathematical conclusion these clumps of matter would form prematurely and collapse almost instantly into black holes, leaving an utterly black universe. So the simplest model of the new inflationary universe, which Hawking and others had helped create, had to be officially pronounced dead – even though it was but six months old.

'Well, at least it showed us the direction we need to be taking, so the conference wasn't a failure,' said Hawking. 'It also showed us that a lot of work remains.'

Still, the inflationary universe is the one cosmologists like the most. 'It appears to be the right approach,' said Hawking. 'It solves more problems than it creates.' One of the products of inflation that theorists are particularly fond of is that an early inflationary burst would have smoothed out all the matter in the early universe to a density that would have allowed that universe to expand as long as ours has.

A rapidly growing young universe could have been so densely packed with matter that it could simply have collapsed back on itself, like a black hole. Or matter might have been spread too thin to bunch into galaxies, and just drifted away into space. The universe's expansion, in fact, required remarkable fine-tuning. This must have occurred within the first 10^{-32} second of the inflationary period.

Although our universe is still in its youth, theorists have a good time exploring its ultimate fate. Do we live in an open universe? Will its current expansion go on for ever, with all matter eventually becoming so diffuse that the stars and galaxies simply blink out one by one? Or do we live in a closed universe? Will the universe one day begin falling back in on itself, reversing the Big Bang in a cataclysmic collapse astrophysicists call the Big Crunch. The universe's fate is still many billions of years off, but nonetheless I asked Hawking his opinion.

'I don't really claim to know the fate of the universe,' he said. 'Nor does anyone else. I think the best guess is that it is just on the borderline between collapse and expansion. But that is just a guess.

'There is a particular cosmological model which predicts a universe expanding with just enough energy to avoid collapse. If I had to pick a model, I think I would pick that one – the one just on the brink of collapse.'

But does not the idea of bubble universes existing side by side suggest that the universe is open, that it will go on expanding into eternity, until it is endlessly dark and cold?

'The idea of numerous universes existing side by side does not affect the idea of an open universe,' answered Hawking. 'It cannot really be considered a classical problem. You would have to take quantum mechanics into consideration – with all its probabilities.

'When you say a number of universes can exist side by side, I think you are treading on dangerous ground – metaphysical ground. I think I am quoting Wittgenstein. He spent half his life in Cambridge. He said – at least I think he said – that the existence of other universes is a predicate. What he meant was that it is not very meaningful to say that other universes exist unless there is some consequence that we can observe.

'In fact, if we can apply quantum mechanics to the universe, then one is led naturally to a picture in which the universe has all sorts of different branches.'

Are these actual, physical regions that could be observed?

'No, these would not be physical branches,' he said. 'It just means that there is a non-zero probability for the universe to have a lot of different forms. Just as there is a probability for it to be open and a probability for it to be closed.

'Maybe', he said, clearly enjoying the speculation, 'it is just that we are in a particular branch of the universe which is on the borderline between being open and being closed. The most remarkable thing about the universe is that it is so close to the borderline between open and closed. The probabilities against it being on such a borderline are enormous. Yet it is still so close that we haven't been able to decide which side it is on.'

Will we ever learn if it is open or closed? Or is that going to turn into another question best left to metaphysics?

'Soon we'll have the equipment,' he said. 'The space telescope should allow us to determine definitely which side of the borderline the universe is on. We'll be able to look much deeper into space and get a more accurate accounting of matter in the universe.

'But it still may be that we can't decide. It is so close. If it is actually sitting on the border in perfect balance, then we'll never know. But, in so far as all our present observations indicate, when the space telescope is in operation we will actually be able to determine if the universe is open or closed.'

'And then what will we know?'

'The fate of the atoms of our bodies.'

I stood on the slopes of Mont Blanc above the ski village of Chamonix in the French Alps. The skies were clear and blue, the mountains shrouded with the first snow of autumn, the leaves just beginning to show their colours. A small Fiat panel truck pulled out of the tunnel that has been cut under Mont Blanc to connect Italy and France. I climbed into it with Roger Antoine, an official at the CERN accelerator laboratory near Geneva. The truck made a U-turn and headed back into the mountain.

Midway through the seven-mile tunnel, the air was rank

with car fumes and the odour of diesel. At that point, in a cave cut out of the rock just off the roadway, a large apparatus has been constructed to test just one thing: does the proton, the most steady and reliable citizen of the universe, last for ever? Or do protons decay like most other particles? Two decades ago the idea of proton decay would have been scientific heresy. Today scientists are taking the notion seriously.

This is because one of the consequences of the various grand unified theories is that protons, previously thought to be immutable, eventually will decay into other particles. The reason, according to theory, is that the strong force holding the proton together and the weak force that causes radioactive decay are ultimately caused by the same basic interaction – one that appeared for just an instant during the first 10^{-32} second of the universe's life. Thus, like a radioactive atom, the proton itself could be doomed to decay – eventually.

The theories predict that, on average, it will take any proton an exceedingly long time – at least 10^{30} years or more – to decay. It was nonetheless a surprisingly easy task for researchers to devise an experiment to test something that will happen far in the future, a timespan so great that it is longer than the age of the universe.

Two miles beneath the summit of Mont Blanc at a location safe from the cosmic radiation that could produce a misleading signal in the apparatus, Picchi Pio, an Italian physicist, showed me an experiment to measure the lifetime of a proton.

'Obviously we can't wait around for billions of years to watch one proton and see if it disappears,' said Pio. 'But we can assemble 10^{30} or more protons and see if one decays during a certain period – say, one year.' If proton decay is a fact of nature, then statistically at least one proton should die during the year.

Pio showed me his collection of protons. They were

contained in an array of stacked iron slabs weighing a total of 150 tons. Pio and his colleagues estimated the number of protons in the iron at nearly 10^{32}, 'just the right number for a good experiment'.

The slabs are fitted throughout with 42,000 devices similar to Geiger counters to pick up the burst of radiation emitted by a dying proton. The devices, called calorimeters, are wired to a computer. The job of Pio and his colleagues from several Italian universities and CERN is to watch – just wait and watch – the computer terminal for the right signal. When I saw the experiment in late 1982, four candidate events had already appeared on the computer screen.

Pio had a computer printout of one. It was a Y-shaped track that the experimenters were reasonably convinced showed that one proton had decayed into a lepton called a muon and a positive electron or positron, leaving in the wake a little burst of energy to be picked up by the calorimeters. Another track he showed me was that of a neutrino he said had passed entirely through the Earth before reaching the tunnel beneath Mont Blanc.

Millions of dollars are being spent on similar experiments in India, Ohio, Minnesota, South Dakota, Utah and Japan to find out the lifetime of protons. So far most of the experiments are inconclusive, although researchers in India claim to have at least eight instances of proton decay. If it is ever shown for certain that protons decay, this will prove that the various GUTs are on the right track. It also will indicate to scientists that the universe is inherently unstable since protons are the major constituents of matter.

Hawking remains sceptical.

'They won't find proton decay,' he told me flatly one day in his office. 'If they do, it will mean there is something wrong with the experiment. My guess is that the lifetime is quite a bit longer than they think. They are looking in the range of 10^{30} to 10^{33} years. That is the best they can do at the moment.

'I would estimate that it is quite a bit longer than 10^{33} years. And, in that case, it will be almost impossible to see.'

'You seem quite certain,' I observed.

'If the very simplest equations in grand unified models are correct, it could be found,' he conceded. 'But one can make grand unified models in which the lifetime would be far longer than 10^{33} years. Then it would never be found.

'For another thing – not only for now, but in the future – it will be impossible to distinguish proton decay from certain other events caused by neutrinos. These events aren't the same, but look very much like proton decay.' I recalled Picchi Pio's neutrino, the one that had made it all the way through the Earth, and its resemblance to the track left by an alleged proton in its death throes.

'There is another kind of proton decay caused by little black holes,' Hawking said. 'These little holes are smaller than protons. But the lifetime of protons in that case is more like 10^{45} years. Nothing could ever measure it.'

Hawking's disdain for the experiments of the proton researchers could be written off as characteristic of the mutual distrust of theorists and experimentalists. He insists not. 'I'm not against the experimenters. Just the methods. But if proton decay were ever proved it would lead to some rather interesting speculation.'

CHAPTER NINE

The Anthropic Principle

Early reports of the Big Bang found a ready audience in many religious groups. After learning what a scientific Genesis entailed, Pope Pius XXII declared in 1951, 'True science to an ever-increasing degree discovers God as though God were waiting behind each door opened by science.' More than a few scientists think that the facts of the Big Bang, as they are slowly uncovered, could at the very least suggest the work of a creator or creative force. It may soon become evident that science will never be able to take us to the exact moment of creation – only up to that point where philosophy, metaphysics and theology begin. Stephen Hawking has made a tentative foray into this uncertain area. 'The odds against a universe like ours emerging out of something like the Big Bang are enormous,' he told me. 'I think there are clearly religious implications whenever you start to discuss the origins of the universe. There must be religious overtones. But I think most scientists prefer to shy away from the religious side of it.'

A few years ago, while thinking about the meaning rather than the numbers of the universe, Hawking and a few colleagues worked out a principle that some scientists considered heresy but others thought put the universe in the right perspective.

Hawking's principle was based on a classic thought experiment. It took as a first premise that all the features of our everyday world, the subatomic world and the cosmos itself are determined by a few basic physical laws and constants, perhaps no more than a total of fifteen. These have been discovered by science and include the masses of the elementary particles and the relative strengths of the basic forces that operate between them.

Hawking, together with Brandon Carter and other colleagues, discovered that an extremely delicate balance exists in nature. For instance, if the strong force that acts on the quarks, neutrons and protons of the atomic nucleus were just slightly weaker, the only element that would be stable would be hydrogen. No other elements could exist.

If the strong force were just a bit stronger in relation to electromagnetism, the force that regulates the way leptons, electrons and neutrinos behave, then an atomic nucleus containing just two protons – a diproton – would become a stable feature of the universe. That would mean that hydrogen would not exist, and the stars and galaxies would have evolved, if at all, far differently from the way they have.

If the constant of gravity were stronger – only 10^{25} times less powerful than the strong nuclear force instead of 10^{38} times weaker, our universe would be small and swift. The average star would have only 10^{-12} times the mass of the sun and could exist for just about a year, hardly time for complex biological phenomena such as mankind to develop.

If gravity were less powerful than it is, then matter would not have congealed into stars and galaxies and the universe would be cold and empty. It is, however, precisely because gravity is so much weaker than the other three forces that our galaxy and solar system evolved. And, as Hawking points out, the growth of the universe – so close to the border between collapse and external expansion that man has not been able to measure it – has been at just the proper rate to allow galaxies and stars to form.

'In fact,' said Hawking, 'a universe like ours with galaxies and stars is actually quite unlikely. If one considers the possible constants and laws that could have emerged, the odds against a universe that has produced life like ours are immense.'

There also is the matter of entropy. This measure of perpetually increasing decay and chaos is controlled by the second law of thermodynamics, which declares that any change in the universe will lead to a slightly more disorderly place. Entropy always goes up. Order always goes down. Evidence of this universal tendency towards disintegration is everywhere. Cars rust, stars grow cold and die, stereos break down, people become old, mountains erode and buildings collapse. This leads to a dilemma: if the universe is a place that is like a watch slowly running down, how, in the face of this natural tendency, did it get wound up in the first place? In defiance of the second law of thermodynamics, order has risen out of chaos.

The second law of thermodynamics is not absolute. Entropy can decrease; that is, order can increase naturally. But it is extremely unlikely. Consider the odds of shaking the parts of a watch in a barrel and having them fall into place as a working timepiece. Is that the kind of event that led to the Big Bang? Is our universe an enormous, accidental reversal of entropy? Or is it – literally – a miracle?

Hawking thinks that the only way to explain our universe is by our presence in it. 'This principle can be paraphrased as "Things are as they are because we are."

'According to one version of the principle, there is a large number of different and separate universes,' he said. 'Each has different values for its physical parameters and for its initial conditions. Most will not have the right conditions for the development of intelligent life.

'However, in a small number there will be conditions and parameters as in our universe. In those, it will be possible for intelligent life to develop and ask the question "Why is the universe as we observe it?" The only answer will be that, if it were otherwise, there would be nobody to ask the question.

'Surprisingly, this principle provides some explanation of many of the remarkable numerical relations that are observed between the values of different physical parameters.'

Brandon Carter calls this rather curious concept the 'anthropic principle'. Some scientists decry Carter and Hawking's anthropic principle on the grounds that it offers no explanation at all. Most theologians have found the principle perplexing and unsatisfactory, since it does not clearly call for the work of a creator.

Nobody has yet been able to show that any universe that emerged out of a flash of creation must possess the features our universe actually possesses. Maybe for now the anthropic principle – a sort of half argument that doesn't really address our curiosity about the origin of the universe – is the best that science can do.

Hawking, the most curious of men, admits that the anthropic principle doesn't come close to providing a true scientific description of the universe in a real sense. 'If we are going to rely on the anthropic principle, we still need some unifying theory to account for the initial conditions of the universe,' he said.

Some physicists take the concept very seriously. John Wheeler at the University of Texas, who has been called the physicist's physicist, has expanded on the anthropic idea and envisions an ensemble of universes in endless cycles of cosmic expansion and contraction. This takes place in an arena he calls 'superspace', an infinite-dimensional space in which every point can correspond to the entire geometry of a universe.

In superspace there is room for nearly every imaginable variety of universe – those that collapse after just a few minutes or those where all stars are green or red. Most of these superspace universes are stillborn in that they are lifeless. Wheeler agrees with Hawking and Carter that our own universe is uniquely fine-tuned to produce life, even if in just one small, lost corner.

In this view, mankind might be the crown jewel of all creation. The universe is the way it is because we have evolved within it. Wheeler even suggests that a universe in which life has failed to evolve is a failed universe. And he has come to believe that a universe constructed so that life did not evolve within it could not have come into existence in the first place.

Wheeler calls this the principle of 'observership'. It is an extension of the quantum idea that without an observer there is no subatomic physics. For Wheeler we live in an observer-dependent, participatory universe. All physical laws are dependent upon the presence of an observer to formulate them.

In fact, he has suggested that this principle leads to the idea that the laws of physics are themselves a counter to the primordial nothing — total entropy. A universe without an observer is not a universe at all.

Recently, some physicists have come to see a relationship between their work and the ideas behind Eastern mysticism. They believe that the paradoxes, odds and probabilities as well as the observer-dependence of quantum mechanics have been anticipated in the writings of Hinduism, Buddhism and Taoism. Quantum mechanics, these so-called new physicists are fond of pointing out, is really only a rediscovery of Shiva or Mahadeva, the Hindu horned god of destruction and cosmic dissolution.

Shiva, mentioned as early as the third or fourth century BC, takes several forms. One of them is Nataraja, the four-armed Lord of the Cosmic Dance pictured dancing on a prostrate demon. The god's dance symbolizes the perpetual process of universal creation and destruction. Matter has no substance at all; it is merely the dynamic, rhythmic gyration of energy coming and going.

David Bohm, Professor of Theoretical Physics at Birkbeck College, is one of these new theoreticians. He thinks the human mind's ability to grasp higher realities is denied or ignored by conventional science. Standard science is a dead end because it analyses experience into discrete pieces. The human mind – and particularly the mind of the physicist – has an overwhelming need to impose categories on experience.

As a result, the seamless web of physical reality is divided into separate events that seem to occur only side by side or in different parts of time and space. By understanding Eastern mysticism, Bohm suggests, physicists can free their minds, at least briefly, from this self-created prison in order to attain an instant of scientific creation.

'The universe of Eastern mysticism is an illusion,' Hawking said. 'A physicist who attempts to link it with his own work has abandoned physics.'

On 29 April 1980, Hawking was inaugurated as Lucasian Professor of Mathematics at Cambridge. The post is one of the university's highest, and his elevation to it was a remarkable achievement.

His inaugural lecture, 'Is the End in Sight for Theoretical Physics?',* was read for him by one of his students.

It was his belief, Hawking said, that soon mankind would have a new theory that would explain what the universe was like at the very beginning and why it behaves today the way it does. This will require a firmer understanding of the four forces observed in nature. The key will be a quantum theory of gravity, which could easily come within twenty years. Hawking ended the lecture on what he called a 'slightly alarmist note'.

'At present, computers are a useful aid in research, but

---

* The complete lecture has been included as an appendix.

they have to be directed by human minds,' he said. 'However, if one extrapolates their recent rapid rate of development, it would seem quite possible that they will take over altogether in theoretical physics. So maybe the end is in sight for theoretical physicists if not for theoretical physics.'

We discussed the lecture two years later. I wondered especially about his closing remarks. 'The point is', he said, 'that we've come such a long way in the last twenty — or fifty — years that one can't hope that it will just go on like that indefinitely. So I think it is altogether possible that we'll either become bogged down and have no more progress or that we'll soon find the unified theory, possibly within twenty more years.'

I asked Hawking about his own future in physics.

'As far as theoretical physics are concerned, I'm already over the hill,' said Hawking. 'Actually, quite far over the hill.' He turned forty-two in January 1984. With the characteristically pragmatic outlook that has typified his battle against stupendous odds for the past two decades, he explained:

'Well, you know most of the best work in theoretical physics is done by people at a very early age — usually by people in their twenties. So being over forty is not a stage in life where one expects to make great discoveries in theoretical physics.'

The reason, he suggested, was that a person begins to lose his mental agility as he ages. 'And young people don't know any better,' he said. 'When they get a radical new idea, they're not afraid to take a chance with it.'

One wonders what keeps Hawking going. Is it stubbornness, his reluctance to take even a single day off when he has the flu or a bad cold? Or is it some other sort of mental toughness, a sort of super-stiff upper lip, that makes Hawking unwilling to complain about, and possibly even unwilling to think about, a condition that would have destroyed lesser men?

It is probably a little bit of both. Stephen Hawking is a very tough man, the toughest man I have ever met. But it goes beyond that. He is our planet's most fully developed cerebral creature, a man who lives to think.

'I think we'll come to the unifying theory within the next two decades, probably in a series of small steps,' he says. 'But, you know, once we find it, it will rather take the fun out of theoretical physics.'

APPENDIX:

———•———

Is the End in Sight for Theoretical Physics?
An Inaugural Lecture

In this lecture I want to discuss the possibility that the goal of theoretical physics might be achieved in the not too distant future, say, by the end of the century. By this I mean that we might have a complete, consistent and unified theory of the physical interactions which would describe all possible observations. Of course, one has to be very cautious about making such predictions. We have thought that we were on the brink of the final synthesis at least twice before. At the beginning of the century it was believed that everything could be understood in terms of continuum mechanics. All that was needed was to measure a certain number of coefficients of elasticity, viscosity, conductivity, etc. This hope was shattered by the discovery of atomic structure and quantum mechanics. Again, in the late 1920s Max Born told a group of scientists visiting Göttingen that 'physics, as we know it, will be over in six months'. This was shortly after the discovery by Paul Dirac, a previous holder of the Lucasian Chair, of the Dirac equation, which governs the behaviour of the electron. It was expected that a similar equation would govern the proton, the only other supposedly elementary particle known at that time. However, the discovery of the neutron and of nuclear forces disappointed these hopes. We now know in fact that neither the proton nor the neutron is elementary but that they are made up of smaller particles.

Nevertheless, we have made a lot of progress in recent years and, as I shall describe, there are some grounds for cautious optimism that we may see a complete theory within the lifetime of some of those present here.

Even if we do achieve a complete unified theory, we shall not be able to make detailed predictions in any but the simplest situations. For example, we already know the physical laws that govern everything that we experience in everyday life. As Dirac pointed out, his equation was the basis of 'most of physics and all of chemistry'. However, we have been able to solve the equation only for the very simplest system, the hydrogen atom consisting of one proton and one electron. For more complicated atoms with more electrons, let alone for molecules with more than one nucleus, we have to resort to approximations and intuitive guesses of doubtful validity. For macroscopic systems consisting of 10^{23} particles or so, we have to use statistical methods and abandon any pretence of solving the equations exactly. Although in principle we know the equations that govern the whole of biology, we have not been able to reduce the study of human behaviour to a branch of applied mathematics.

What would we mean by a complete and unified theory of physics? Our attempts at modelling physical reality normally consist of two parts:

1. A set of local laws that are obeyed by the various physical quantities. These are usually formulated in terms of differential equations.

2. Sets of boundary conditions that tell us the state of some regions of the universe at a certain time and what effects propagate into it subsequently from the rest of the universe.

Many people would claim that the role of science was confined to the first of these and that theoretical physics will have achieved its goal when we have obtained a complete set of local physical laws. They would regard the question of the initial conditions for the universe as belonging to the realm of metaphysics or religion. In a way this attitude is similar to that of those who in earlier centuries discouraged scientific investigation by saying that all natural phenomena were the work of God and should not be inquired into. I think that the initial conditions of the universe are as suitable a subject for scientific study and theory as are the local physical laws. We shall not have a complete theory until we can do more than merely say that 'things are as they are because they were as they were'.

The question of the uniqueness of the initial conditions is closely related to that of the arbitrariness of the local physical laws: one would not regard a theory as complete if it contained a number of adjustable parameters such as masses or coupling constants which could be given any values one liked. In fact it seems that neither the initial conditions nor the values of the parameters in the theory are arbitrary but that they are somehow chosen or picked out very carefully. For example, if the proton–neutron mass difference were not about twice the mass of the electron, one would not obtain the couple of hundred or so stable nucleides that make up the elements and are the basis of chemistry and biology. Similarly, if the gravitational mass of the proton were significantly different, one would not have had stars in which these nucleides could have been built up and, if the initial expansion of the universe had been slightly smaller or greater, the universe would either have collapsed before such stars could have evolved or would have expanded so rapidly that stars would never have been formed by gravitational condensation. Indeed, some people have gone so far as to elevate these restrictions on the initial conditions and the parameters to the status of a principle, the anthropic principle, which can be paraphrased as 'Things are as they are because we are.' According to one version of the principle, there is a very large number of different separate universes with different values of the physical parameters and different initial conditions. Most of these universes will not provide the right conditions for the development of the complicated structures needed for intelligent life. Only in a small number, with conditions and parameters like our own universe, will it be possible for intelligent life to develop and to ask the question 'Why is the universe as we observe it?' The answer is, of course, that if it were otherwise there would not be anyone to ask the question.

The anthropic principle does provide some sort of explanation of many of the remarkable numerical relations that are observed between the values of different physical parameters. However, it is not completely satisfactory; one cannot help feeling that there is some deeper explanation. Also, it cannot account for all the regions of the universe. For example, our solar system is certainly a prerequisite for our existence as is an earlier generation of nearby stars in which heavy elements could have been formed by nuclear synthesis. It might even be that the whole of our galaxy was required. But there does not seem any necessity for other galaxies to exist, let alone the million million or so of them that we see, distributed roughly uniformly throughout the observable universe. This large-scale homogeneity of the universe makes it very difficult to hold an anthropocentric view or to believe that the structure of the universe is determined by anything so peripheral as some complicated molecular structures on a minor planet orbiting a very average star in the outer suburbs of a fairly typical spiral galaxy.

If we are not going to appeal to the anthropic principle, we need some unifying theory to account for the initial conditions of the universe and the values of the various physical parameters. However, it is too difficult to think up a complete theory of everything all at one go (though this does not seem to stop some people; I get two or three unified theories in the mail each week). What we do instead is to look for partial theories that will describe situations in which certain interactions can be ignored or approximated in a simple manner. We first divide the material content of the universe into two parts: 'matter', particles such as quarks, electrons, muons, etc; and 'interactions' such as gravity, electromagnetism, etc. The matter particles are described by fields of one-half-integer spin and obey the Pauli Exclusion Principle, which prevents more than one particle of a given kind from being in any state. This is the reason that we can have solid bodies that do not collapse to a point or radiate away to infinity. The matter principles are divided into two groups: the hadrons, which are composed of quarks; and the leptons, which comprise the remainder.

The interactions are divided phenomenologically into four categories. In order of strength they are: the strong nuclear forces, which interact only with hadrons; electromagnetism, which interacts with charged hadrons and leptons; the weak nuclear forces, which interact with all hadrons and leptons; and finally, the weakest by far, gravity, which interacts with everything. The interactions are represented by integer-spin fields which do not obey the Pauli Exclusion Principle. This means that they can have many particles in the same state. In the case of electromagnetism and gravity, the interactions are also long-range, which means that the fields produced by a large number of matter particles can all add up to give a field that can be detected on a macroscopic scale. For these reasons they were the first to have theories developed for them, gravity by Newton in the seventeenth century and electromagnetism by Maxwell in the nineteenth century. However, these theories were basically incompatible because the Newtonian theory was invariant if the whole system was given any uniform velocity, whereas the Maxwell theory defined a preferred velocity, the speed of light. In the end it turned out to be the Newtonian theory of gravity which had to be modified to make it compatible with the invariance properties of the Maxwell theory. This was achieved by Einstein's general theory of relativity, which was formulated in 1915.

The general relativity theory of gravity and the Maxwell theory of electrodynamics were what is called classical theories, that is, they involved quantities which were continuously variable and which could, in principle at least, be measured to arbitrary accuracy. However, a problem arose when one tried to use such theories to construct a model of the atom. It had been discovered that the atom consisted of a small positively charged nucleus surrounded by a cloud of negatively charged electrons. The natural assumption was that the electrons were in orbit around the nucleus as the Earth is in orbit around the Sun. However, the classical theory predicted that the electrons would radiate electromagnetic waves. These waves would carry away energy and would cause the electrons to spiral into the nucleus, producing a collapse of the atom.

This problem was overcome by what is undoubtedly the greatest achievement in theoretical physics this century, the discovery of the quantum theory. The basic postulate of this is the Heisenberg Uncertainty Principle, which states that certain pairs of quantities, such as the position and momentum of a particle, cannot be measured simultaneously with arbitrary accuracy. In the case of the atom this meant that in its lowest energy state the electron could not be at rest in the nucleus because, in that case, its position and velocity would both be defined exactly. Instead, the electron would have to be smeared out with some probability distribution around the nucleus. In this state the electron could not radiate energy in the form of electromagnetic waves because there would be no lower energy state for it to go to.

In the 1920s and 1930s quantum mechanics was applied with great success to systems such as atoms or molecules, which have only a finite number of degrees of freedom. Difficulties arose, however, when people tried to apply it to the electromagnetic field, which has an infinite number of degrees of freedom, roughly speaking two for each point of space-time. One can regard these degrees of freedom as oscillators, each with its own position and momentum. The oscillators cannot be at rest because then they would have exactly defined positions and momenta. Instead, each oscillator must have some minimum amount of what are called 'zero-point fluctuations' and a non-zero energy. The energies of the zero-point fluctuations of all the infinite number of degrees of freedom would cause the apparent mass and charge of the electron to become infinite.

A procedure called renormalization was developed to overcome this difficulty in the late 1940s. It consisted of the rather arbitrary subtraction of certain infinite quantities to leave finite remainders. In the case of electrodynamics, it was necessary to make two such infinite subtractions, one for the mass and the other for the charge of the electron. This renormalization procedure has never been put on a very firm conceptual or mathematical basis, but it has worked quite well in practice. Its great success was the prediction of a small displacement, the Lamb shift, in some lines in the spectrum of atomic hydrogen. However, it is not very satisfactory from the point of view of attempts to construct a complete theory because it does not make any predictions of the values of the finite remainders left after making infinite subtractions. Thus, we would have to fall back on the anthropic principle to explain why the electron has the mass and charge that it does.

During the 1950s and 1960s it was generally believed that the weak and strong nuclear forces were not renormalizable; that is, they would require an infinite number of infinite subtractions to make them finite. There would be an infinite number of finite remainders which were not determined by the theory. Such a theory would have no predictive power because one could never measure all the infinite number of parameters. However, in 1971 'tHooft showed that a unified model of the electromagnetic and weak interactions that had been earlier proposed by Salam and Weinberg was indeed renormalizable with only a finite number of infinite subtractions. In the Salam–Weinberg theory the photon, the spin-1 particle that carries the electromagnetic interaction, is joined by three other spin-1 partners called W^+, W^- and Z°. At very high energies these four particles are all predicted to behave in a similar manner. However, at lower energies a phenomenon called 'spontaneous symmetry breaking' is invoked to explain the fact that the photon has zero rest mass whereas the W^+, W^- and Z° are all very massive. The low energy predictions of this theory have agreed remarkably well with observation, and this led the Swedish Academy last year to award the Nobel Prize to Salam, Weinberg and Glashow, who had also constructed similar unified theories. However, Glashow himself remarked that the Nobel Committee really took rather a gamble because we do not yet have particle accelerators of high enough energy to test the theory in the regime where unification between the electromagnetic forces, carried by the photon, and the weak forces, carried by the W^+, W^- and Z°, really occurs. Sufficiently powerful accelerators will be ready in a few years, and most physicists are confident that they will confirm the Salam–Weinberg theory.

The success of the Salam–Weinberg theory led to the search for a similar renormalizable theory of the strong interactions. It was realized fairly early on that the proton and other hadrons such as the pi meson could not be truly elementary particles, but that they must be bound states of other particles called quarks. These seem to have the curious property that, although they can move fairly freely within a hadron, it appears to be impossible to obtain just one quark on its own; they always come either in groups of three (like the proton or neutron) or in pairs consisting of a quark and antiquark (like the pi meson). To explain this, quarks were endowed with an attribute called colour. It should be emphasized that this has nothing to do with our normal perception of colour; quarks are far too small to be seen by visible light. It is merely a convenient name. The idea is that quarks come in three colours – red, green and blue – but that any isolated bound state such as a hadron has to be colourless, either a combination of red, green and blue like the proton or a mixture of red and anti-red, green and anti-green, and blue and anti-blue like the pi meson.

The strong interactions between the quarks are supposed to be carried by spin-1 particles called gluons, rather like the particles that carry the weak interaction. The gluons also carry colour, and they and the quarks obey a renormalizable theory called quantum chromodynamics, or QCD for short. A consequence of the renormalization procedure is that the effective coupling constant of the theory depends on the energy at which it is measured and decreases to zero at very high energies. This phenomenon is known as asymptotic freedom. It means that quarks inside a hadron behave almost like free particles in high-energy collisions so that their interactions can be treated successfully by perturbation theory. The predictions of perturbation theory are in reasonable qualitative agreement with observation, but one cannot yet really claim that the theory has been experimentally verified. At low energies the effective coupling constant becomes very large and perturbation theory breaks down. It is hoped that this 'infra-red slavery' will explain why quarks are always confined in colourless bound states, but so far no one has been able to demonstrate this really convincingly.

Having obtained one renormalizable theory for the strong interactions and another one for the weak and electromagnetic interactions, it was natural to look for a theory which combined the two. Such theories are given the rather exaggerated title of 'grand unified theories' or GUTs. This is rather misleading because they are neither all that grand, nor fully unified, nor complete theories in that they have a number of undetermined renormalization parameters such as coupling constants and masses. Nevertheless they may be a significant step towards a complete unified theory. The basic idea is that the effective coupling constant of the strong interactions, which is large at low energies, gradually decreases at high energies because of asymptotic freedom. On the other hand, the effective coupling constant of the Salam–Weinberg theory, which is small at low energies, gradually increases at high energies because this theory is not asymptotically free. If one extrapolates the low-energy rate of increase and decrease of the coupling constants, one finds that the two coupling constants become equal at an energy of about 10^{15} GeV. The theories propose that above this energy the strong interactions are unified with the weak and electromagnetic interactions but that at lower energies there is spontaneous symmetry breaking.

An energy of 10^{15} GeV is way beyond the scope of any laboratory equipment: the present generation of particle accelerators can produce centre-of-mass energies of about 10 GeV, and the next generation will produce energies of 100 GeV or so. This will be sufficient to investigate the energy range in which the electromagnetic forces should become unified with the weak forces according to the Salam–Weinberg theory, but not the enormously high energy at which the weak and electromagnetic interactions would be predicted to become unified with the strong interactions. Nevertheless there can be low-energy predictions of the grand unified theories that might be testable in

118

the laboratory. For example, the theories predict that the proton should not be completely stable but should decay with a lifetime of order 10^{31} years. The present experimental lower limit on the lifetime is about 10^{30} years, and it should be possible to improve this.

Another observable prediction concerns the ratio of baryons to photons in the universe. The law of physics seems to be the same for particles and antiparticles. More precisely, they are the same if particles are replaced by antiparticles, right-handed is replaced by left-handed, and the velocities of all particles are reversed. This is known as the CPT theorem, and it is a consequence of basic assumptions that should hold in any reasonable theory. Yet the Earth and indeed the whole solar system is made up of protons and neutrons without any antiprotons or antineutrons. Indeed, such an imbalance between particles and antiparticles is yet another *a priori* condition for our existence, for if the solar system were composed of an equal mixture of particles and antiparticles they would all annihilate each other and leave just radiation. From the observed absence of such annihilation radiation we can conclude that our galaxy is made entirely of particles rather than antiparticles. We do not have direct evidence about other galaxies, but it seems likely that they are composed of particles and that in the universe as a whole there is an excess of particles over antiparticles of about one particle per 10^8 photons. One could try to account for this by invoking the anthropic principle, but grand unified theories actually provide a possible mechanism for explaining the imbalance. Although all interactions seem to be invariant under the combination of C (replace particles by antiparticles), P (change right-handed to left-handed) and T (reverse the direction of time), there are known to be interactions which are not invariant under T alone. In the early universe, in which there is a very marked arrow of time given by the expansion, these interactions could produce more particles than antiparticles. However, the number they make is very model-dependent, so that agreement with observation is hardly a confirmation of the grand unified theories.

So far most of the effort has been devoted to unifying the first three categories of physical interactions, the strong and weak nuclear forces and electromagnetism. The fourth and last, gravity, has been neglected. One justification for this is that gravity is so weak that quantum gravitational effects would be large only at particle energies way beyond those in any particle accelerator. Another is that gravity does not seem to be renormalizable: in order to obtain finite answers it seems that one may have to make an infinite number of infinite subtractions with a correspondingly infinite number of undetermined finite remainders. Yet one must include gravity if one is to obtain a fully unified theory. Furthermore the classical theory of general relativity predicts that there should be space-time singularities at which the gravitational field would become infinitely strong. These singularities would occur in the past at the beginning of the present expansion of the universe (the Big Bang) and in the future in the gravitational collapse of stars and, possibly, of the universe itself. The prediction of singularities presumably indicates that the classical theory will break down. However, there seems to be no reason why it should break down until the gravitational field becomes strong enough so that quantum gravitational effects are important. Thus a quantum theory of gravity is essential if we are to describe the early universe and then give some explanation for the initial conditions beyond merely appealing to the anthropic principle.

Such a theory is also required if we are to answer the question, Does time really have a beginning and, possibly, an end as is predicted by classical general relativity, or are the singularities in the Big Bang and the Big Crunch smeared out in some way by quantum effects? This is a difficult question to give a well-defined meaning to when the very structure of space and time themselves are subject to the uncertainty principle. My personal feeling is that singularities are probably still present, though one can continue time past them in a certain mathematical sense. However, any subjective concept of time that was related to consciousness or the ability to perform measurements would come to an end.

What are the prospects of obtaining a quantum theory of gravity and of unifying it with the other three categories of interactions? The best hope seems to lie in an extension of general relativity called supergravity. In this the graviton, the spin-2 particle that carries the gravitational interaction, is related to a number of other fields of lower spin by so-called supersymmetry transformations. Such a theory has the greater merit that it does away with the old dichotomy between 'matter', represented by particles of one-half-integer spin, and 'interactions', represented by integer-spin particles. It also has the great advantage that many of the infinities which arise in quantum theory cancel each other out. Whether or not they all cancel out to give a theory which is finite without any infinite subtractions is not yet known. It is hoped that they do because it can be shown that theories which include gravity are either finite or nonrenormalizable; that is, if one has to make any infinite subtractions, then one will have to make an infinite number of them with a corresponding infinite number of undetermined remainders. Thus, if all the infinities in supergravity turn out to cancel each other out, we could have a theory which not only fully unified all the matter particles and interactions, but which was also complete in the sense that it did not have any undetermined renormalization parameters.

Although we do not yet have a proper quantum theory of gravity, let alone one which unifies it with the other physical interactions, we have an idea of some of the features it should have. One of these is connected with the fact that gravity affects the casual structure of space-time; that is, gravity determines which events can be causally related to each other. An example of this in the classical theory of general relativity is provided by a black hole, which is a region of space-time in which the gravitational field is so strong that any light or other signal is dragged back into the region and cannot escape to the outside world. The intense gravitational field near the black hole causes the creation of pairs of particles and antiparticles, one of which falls into the black hole and the other of which escapes to infinity. The particle that escapes appears to have been emitted by the black hole. An observer at a distance from the black hole can measure only the outgoing particles, and he cannot correlate them with those that fell into the hole because he cannot observe them. This means that the outgoing particles have an extra degree of randomness or unpredictability over and above that usually associated with the uncertainty principle. In normal situations the uncertainty principle implies that one can definitely predict *either* the position *or* the velocity of a particle *or* one combination of position and velocity. Thus, roughly speaking, one's ability to make definite predictions is halved. However, in the case of particles emitted from a black hole, the fact that one cannot observe what is going on inside the black hole means that one can definitely predict *neither* the positions *nor* the velocities of the emitted particles. All one can give are probabilities that particles will be emitted in certain modes.

It seems, therefore, that even if we find a unified theory we may be able to make only statistical predictions. We would also have to abandon the view that there is a unique universe that we observe. Instead, we would have to adopt a picture in which there was an ensemble of all possible universes with some probability distribution. This might explain why the universe started off in the Big Bang in almost perfect thermal equilibrium, because thermal equilibrium would correspond to the largest number of microscopic configurations and hence the great probability. To echo Voltaire's philosopher Pangloss, 'We live in the most probable of all possible worlds.'

What are the prospects that we will find a complete unified theory in the not too distant future? Each time we have extended our observations to small length scales and higher energies, we have discovered new layers of structure. At the beginning of the century, the discovery of Brownian motion with a typical energy particle of 3×10^{-2} eV showed that matter was not continuous but was made up of atoms. Shortly thereafter it was discovered that these supposedly indivisible atoms were made up of electrons revolving about a nucleus with energies of the order of a few electron-volts. The nucleus in turn was found to be composed of so-called elementary particles, protons and neutrons, held together by nuclear bonds of the order of 10^6 eV. The latest episode in this story is that we have found that the proton and the neutron are made up of quarks held together by bonds of order 10^9 eV. It is a tribute to how far we have come already in theoretical physics that it now takes enormous machines and a great deal of money to perform an experiment whose results we cannot predict.

Our past experience might suggest that there is an infinite sequence of layers of structure at higher and higher energies. Indeed, such a view of an infinite regress of boxes within boxes was official dogma in China under the Gang of Four. However, it seems that gravity should provide a limit but only at the very short length scale of 10^{-33} cm or the very high energy of 10^{28} eV. On length scales shorter than this, one would expect that space-time would cease to behave like a smooth continuum and that it would acquire a foamlike structure because of quantum fluctuations of the gravitational field.

There is a very large unexplored region between our present experimental limit of about 10^{10} eV and the gravitational cutoff at 10^{28} eV. It might seem naïve to assume, as grand unified theories do, that there are only one or two layers of structure in this enormous interval. However, there are grounds for optimism. At the moment, at least, it seems that gravity can be unified with the other physical interactions only in some supergravity theory. There appear to be only a finite number of such theories. In particular, there is a largest such theory, the so-called $N = 8$ extended supergravity. This contains one graviton, eight spin-$3/2$ particles called gravitonos, twenty-eight spin-1 particles, fifty-six spin-$1/2$ particles, and seventy particles of spin O. Large as these numbers are, they are not large enough to account for all the particles that we seem to observe in strong and weak interactions. For instance, the $N = 8$ theory has twenty-eight spin-1 particles. These are sufficient to account for the gluons that carry the strong interactions and two of the four particles that carry the weak interactions but not the other two. One would therefore have to believe that many or most of the observed particles such as gluons or quarks are not really elementary as they seem at the moment but that they are bound states of the fundamental $N = 8$ particles. It is not likely that we shall have accelerators powerful enough to probe these composite structures within the foreseeable future, or indeed ever, especially if one makes a projection based on the current economic trends. Nevertheless, the fact that these bound states arose from the well-defined $N = 8$ theory should enable us to make a number of predictions that could be tested at energies that are accessible now or will be in the near future. The situation might thus be similar to that for the Salam–Weinberg theory unifying electromagnetism and weak interactions. The low-energy predictions of this theory are in such good agreement with observation that the theory is now generally accepted even though we have not yet reached the energy at which the unification should take place.

There ought to be something very distinctive about the theory that describes the universe. Why does this theory come to life while other theories exist only in the minds of their inventors? The $N = 8$ supergravity theory does have some claims to be special. It seems that it may be the only theory

1. which is in four dimensions
2. which incorporates gravity
3. which is finite without any infinite subtractions.

I have already pointed out that the third property is necessary if we are to have a complete theory without parameters. It is, however, difficult to account for properties 1 and 2 without appealing to the anthropic principle. There seems to be a consistent theory which satisfies properties 1 and 3 but which does not include gravity. However, in such a universe there would probably not be sufficient in the way of attractive forces to gather together matter into the large aggregates which are probably necessary for the development of complicated structures. Why space-time should be four-dimensional is a question that is normally considered to be outside the realm of physics. However, there is a good anthropic principle argument for that, too. Three space-time dimensions – that is, two space and one time – are clearly insufficient for any complicated organism. On the other hand, if there were more than three spatial dimensions, the orbits of planets around the sun or electrons around a nucleus would be unstable and they would tend to spiral inward. There remains the possibility of more than one time dimension, but I for one find such a universe very hard to imagine.

So far I have implicitly assumed that there is an ultimate theory. But is there? There are at least three possibilities:

1. There is a complete unified theory.
2. There is no ultimate theory, but there is an infinite sequence of theories which are such that any particular class of observations can be predicted by taking a theory sufficiently far down the chain.
3. There is no theory. Observations cannot be described or predicted beyond a certain point but are just arbitrary.

The third view was advanced as an argument against the scientists of the seventeenth and eighteenth centuries. 'How could they formulate laws which would curtail the freedom of God to change his mind?' Nevertheless they did, and got away with it. In modern times we have effectively eliminated possibility 3 by incorporating it within our scheme: quantum mechanics is essentially a theory of what we do not know and cannot predict.

Possibility 2 would amount to the picture of an infinite sequence of structures at higher and higher energies. As I said before, this seems unlikely because one would expect that there would be a cutoff at the Planck energy of 10^{28} eV. This leaves us with possibility 1. At the moment the N = 8 supergravity theory is the only candidate in sight. There are likely to be a number of crucial calculations within the next few years which have the possibility of showing that the theory is no good. If the theory survives these tests, it will probably be some years more before we develop computational methods that will enable us to make predictions and before we can account for the initial conditions of the universe as well as the local physical laws. These will be the outstanding problems for theoretical physicists in the next twenty years or so. But, to end on a slightly alarmist note, they may not have much more time than that. At present, computers are a useful aid in research, but they have to be directed by human minds. However, if one extrapolates their recent rapid rate of development, it would seem quite possible that they will take over altogether in theoretical physics. So maybe the end is in sight for theoretical physicists if not for theoretical physics.